State of the Art in Separation Science

State of the Art in Separation Science

Editor

Victoria Samanidou

MDPI • Basel • Beijing • Wuhan • Barcelona • Belgrade • Manchester • Tokyo • Cluj • Tianjin

Editor
Victoria Samanidou
Aristotle University of Thessaloniki
Greece

Editorial Office
MDPI
St. Alban-Anlage 66
4052 Basel, Switzerland

This is a reprint of articles from the Topical Collection published online in the open access journal *Separations* (ISSN 2297-8739) (available at: https://www.mdpi.com/journal/separations/topical_collections/Separation_Science).

For citation purposes, cite each article independently as indicated on the article page online and as indicated below:

LastName, A.A.; LastName, B.B.; LastName, C.C. Article Title. *Journal Name* **Year**, *Volume Number*, Page Range.

ISBN 978-3-0365-5385-6 (Hbk)
ISBN 978-3-0365-5386-3 (PDF)

Cover image courtesy of Victoria Samanidou

Contents

About the Editor

Victoria Samanidou

Dr Victoria Samanidou is a Full Professor and Director of the Laboratory of Analytical Chemistry in the Department of Chemistry of Aristotle University of Thessaloniki, Greece.

Her research interests focus on the development of sample preparation methods using sorptive extraction prior to chromatographic analysis. She has co-authored 202 original research articles in peer-reviewed journals and 60 reviews, 66 editorials/in view and 52 chapters in scientific books (h-index 42). She is an editorial board member of more than 30 scientific journals and a guest editor of more than 30 Special Issues. She has peer reviewed more than 700 manuscripts for 149 scientific journals.

In 2016, she was included in top 50 power list of women in Analytical Science, as proposed by Texere Publishers (https://theanalyticalscientist.com/power-list/the-power-list-2016).

In 2021, she was included in the "The Analytical Scientist" 2021 Power List of top 100 influential people in analytical science (https://theanalyticalscientist.com/power-list/2021).

In 2020, she was included in the top 2% of scientists in the world in the field of Analytical Chemistry (career—2019; single year 2019). In 2021, she was again included in the list of World Top 2% Scientists for her whole career up to now, as well as for the single year of 2020.

She is Leader of Working Group 1 Science and Fundamentals of EuChemS-DAC SamplePreparation Study Group and Network since 2021 (https://www.sampleprep.tuc.gr/en/working-groups/wg1-science-and-fundamentals).

In 2016 she was elected as President of the Steering Committee of the Association of Greek Chemists-Regional Division of Central and Western Macedonia.

Preface to "State of the Art in Separation Science"

The term "separation techniques" refers to a group of analytical techniques used to separate the components of a homogenous mixture that are further detected, identified and quantitatively determined.

Though separation can involve distillation and evaporation, typically in analytical chemistry, separation techniques include all types of chromatographic and electrophoretic techniques, such as:

Gas Chromatography (GC); Liquid Chromatography (LC); High-Performance Liquid Chromatography (HPLC); LC-MS; LC-MS/MS; LC-NMR; 2D LC; 2D GC; GC-MS; Capillary Electrochromatography (CEC); Supercritical Fluid Chromatography (SFC); Affinity Electrophoresis; Capillary Electrophoresis; Gel Electrophoresis; Immunoelectrophoresis; Isoelectric Focusing ((IEF); Isotachophoresis (ITP).

Significant advances have been made in the last few decades, including in instrumentation as well as the materials used in stationary or mobile phases. All these advances are following the principles of Green Analytical Chemistry. Therefore, automation, speed and safety are of great importance.

Two-dimensional techniques, as well as those hyphenated with spectroscopic techniques, have improved the feature characteristics of established separation techniques, making them the best analytical tool in the analysis of any kind of sample derived from various chemistry sectors as well other scientific fields such as food science, veterinary medicine, biology, dentistry, archeology, forensics, pharmaceutical chemistry, medicine, environmental science, etc. To that end, major advances in separation science have led to significant findings in the above-mentioned scientific fields.

The questions that arise are: What is the state-of-the-art in Separation Sciences? What advances have been reported recently? Last but not least, what are the future perspectives?

This Topical Collection aims to gather the answers to these questions addressed by the experts in the field who are invited to contribute with a research or review article.

The Editor wishes to thank all authors for their valuable contribution.

Victoria Samanidou
Editor

separations

MDPI

Review

State of the Art in Separation Processes for Alternative Working Fluids in Clean and Efficient Power Generation

Odi Fawwaz Alrebei [1,*], Abdulkarem I. Amhamed [1], Muftah H. El-Naas [2], Mahmoud Hayajnh [3,4], Yasmeen A. Orabi [5], Ward Fawaz [2], Ahmad S. Al-tawaha [4,6] and Agustin Valera Medina [7]

1 Qatar Environment and Energy Research Institute, Hamad Bin Khalifa University, Doha P.O. Box 34110, Qatar; aamhamed@hbku.edu.qa
2 Gas Processing Center, College of Engineering, Qatar University, Doha P.O. Box 2713, Qatar; muftah@qu.edu.qa (M.H.E.-N.); fawazodi@yahoo.com (W.F.)
3 School of Aerospace Engineering, Georgia Institute of Technology, North Avenue NW, Atlanta, GA 30332, USA; mhayajnh@gatech.edu
4 Department of Aeronautical Engineering, Jordan University of Science and Technology, Ramtha 3030, Jordan; Ahmad.s.Tawaha@gmail.com
5 Department of Chemical Engineering, Jordan University of Science and Technology, Ramtha 3030, Jordan; Yorabi89@gmail.com
6 Department of Electrical & Computer Engineering, Virginia Polytechnic Institute and State University, Blacksburg, VA 24061, USA
7 Mechanical and Aerospace Engineering Department, Cardiff School of Engineering, Queen's Buildings Cardiff University, Cardiff CF24 0EA, UK; valeramedinaa1@cardiff.ac.uk
* Correspondence: alrebeio@outlook.com

Abstract: Gas turbines must now comply with much stricter emission control regulations. In fact, to combat the greenhouse effect, regulatory authorities have drastically reduced allowable emission levels. For example, in less than 12 years, the United States' Clean Air Act issued the New Source Performance Standards (NSPS), which tightened the NOx emission margin of natural gas combustion (from 75 ppm to 10 ppm). On the other hand, despite those efforts, the high demand for energy produced by fossil-fueled gas turbines in power plants has resulted in dramatic increases in anthropogenic CO_2 and NO_x emitted by gas combustors. Most systems responsible for these undesirable emissions are directly linked to power generation, with gas turbines playing a pivotal role. Yet, gas turbines are still widely used in power plants and will continue to meet the growing demand. Therefore, sequestration and separation techniques such as Carbon Capture and Storage (CCS) and Air Separation Units (ASU) are essential to reduce CO_2 and NO_x emissions while allowing large amounts of power to be generated from these systems. This paper provides an in-depth examination of the current state of the art in alternative working fluids utilized in the power generation industry (i.e., gas turbines, combustion). In addition, this paper highlights the recent contribution of integrating separation techniques, such as air separation, steam methane reforming, and water-gas shifting, to the power generation industry to facilitate a continuous and adequate supply of alternative working fluids.

Keywords: air separation; steam methane reforming; water gas shift; alternative fluids; gas turbine

Citation: Alrebei, O.F.; Amhamed, A.I.; El-Naas, M.H.; Hayajnh, M.; Orabi, Y.A.; Fawaz, W.; Al-tawaha, A.S.; Medina, A.V. State of the Art in Separation Processes for Alternative Working Fluids in Clean and Efficient Power Generation. *Separations* **2022**, *9*, 14. https://doi.org/10.3390/separations9010014

Academic Editor: Victoria Samanidou

Received: 11 November 2021
Accepted: 7 January 2022
Published: 13 January 2022

1. Introduction

This paper examines the state of art in integrating CCS and separation units to a variety of gas turbine cycle configurations, highlighting their role in producing a continuous and sustainable supply of complex blends of working fluids, thus reducing carbon and NO_x emissions and increasing the power generation cycle efficiency.

Although the progress of separation processes has been reported previously, this paper aims to correlate this progress with the potential of producing a continuous and sustainable supply of alternative working fluids to gas turbine and combustion chambers, hence delivering unique state-of-the-art evaluation. The method of evaluating the involved literature in this paper is based on providing a pros–cons evaluation of each separation process and efficiency comparison tables for each aspect in reference to its potential to produce a continuous and sustainable supply of alternative working fluids.

The literature [1–5] is enriched with research on the use of alternative working fluids, which have significant potential for improving gas turbine efficiency and power outputs. These increases are necessary to ensure the integration of other high-energy-consuming processes into the global carbon footprint reduction effort. Helium in the nuclear industry, ammonia–water in organic Rankine cycles, and humidified injection techniques are all examples of alternative working fluids [5–10].

Alternative working fluids are based on replacing air with an oxygen-based mixture to increase the cycle efficiency and reduce the greenhouse effect. By replacing air with oxygen-based blends, the N_2 content is eliminated from the combustion process; thus, this technique eliminates NO_x emissions while a cost-effective and straightforward condensation process captures carbon emissions [11–16]. The captured carbon is typically circulated back to serve as a diluting agent to avoid turbine overheating associated with oxyfuel combustion [17–20]. As far as efficiency is concerned, the cycle efficiency can be increased by increasing the overall heat capacities (i.e., specific heat at constant pressure (C_p)) of the working fluid in the expansion stage (i.e., turbine). This is achieved by injecting components with high heat capacities (i.e., inert gases) into the working fluid, thus increasing the work output as described in Equation (1). Another approach to improving the work output is to increase the mass flow rate ($\dot{m}$)in the expansion stage by injecting steam into the working fluid as described in Equation (1), where W_t, $\dot{m}$, C_p, T_{03}, and T_{04} are the turbine work, working fluid mass flow rate, isobaric mass heat capacity, turbine total inlet temperature, and turbine total outlet temperature, respectively.

$$W_t = \dot{m}\, C_p(T_{03} - T_{04}) \tag{1}$$

As NO_x emissions are eliminated from the combustion process, alternative working fluid-based gas turbines utilize diffusive combustors due to their high combustion stabilities compared to premixed combustors.

Because of their non-reactive nature combined with their thermodynamic properties, inert gases present a unique opportunity among the potential gases that can be used to increase power and efficiency. Because of its stable radioactive properties, helium is known to be the best choice in high-temperature gas-cooled nuclear reactors (HTGR) [21,22]. Efficiency losses, on the other hand, are the cost of making necessary changes to the compressor and turbine blade geometry. This results in forming an end-wall boundary layer and secondary flow, impacting the final efficiency [23]. In the literature, attempts have been made to compromise between mechanical requirements and high efficiency by introducing inert gases into the working fluid, such as neon (Ne) and helium (He). For example, by optimizing the compositions of both components, an optimal expansion ratio can be achieved.

Nonetheless, such a composition has no discernible impact on cycle efficiency. As a result, additional gases are required to improve overall efficiency, lowering maintenance and operating costs. Furthermore, to accommodate carbon sequestration at the end of the line, the gas blend should include CO_2.

Oxyfuel combustion is one method for using carbon dioxide in carbon sequestration systems that is currently being developed. Oxyfuel combustion is a modern technique that substitutes high oxygen concentrations for air (i.e., with a 21 percent v. oxygen content) as the primary oxidant [24]. Compared to air–fuel combustion, oxyfuel combustion produces flue gases with approximately 75% lower mass and volume (primarily, steam, and CO_2). Compared to air–fuel combustion, CO_2 is captured and circulated back to the oxyfuel combustion-based gas turbine with lower heat losses, resulting in lower efficiency losses [25].

Using CO_2 without injecting inert gases in oxyfuel combustion systems, on the other hand, lowers the turbine inlet temperature, necessitating higher pressure ratios to maintain the same temperature. Essentially, this entails the use of larger compressors, which impact the gas turbine's compactness. Furthermore, existing gas turbine arrangements would necessitate significant changes, delaying the adoption of CO_2 injection [25,26]. As a solution, injecting argon (Ar) into the working fluid increases the overall heat capacity by introducing another inert gas into the working fluid. Injecting Ar is a more economically sustainable option than He, Ne, or other inert gases because of its abundance in the atmosphere. By injecting argon, the turbine inlet temperature can be maintained at a constant level while maintaining high specific heat ratios for improved cycle efficiency. As a result, significant changes to existing gas turbine configurations can be avoided. Furthermore, argon's high heat capacity boosts power outputs, offsetting the efficiency reduction penalty associated with CCS installations.

However, if only Ar-O_2-CO_2 blends were used as working fluids, the costs would remain significant. Therefore, advanced humidified systems could be an alternative to increase output power and efficiencies while lowering the overall costs. Complex cycles are defined as those that combine heat recovery (HR) with humidified injection techniques [27]. By recovering about 60% of heat losses, these cycles improve efficiency and reduce emissions [28]. Heat exchange and recuperation techniques, such as gas-to-gas recuperation, are examples of heat recovery methods for gas turbines with low-pressure ratios [29]. Furthermore, higher thermal efficiencies and output power are demonstrated when heat-exchanged gas turbines are compared to similar simple gas turbines. On the other hand, steam injection is a more preferable approach for higher pressure ratios [29].

As far as the evaporative cycles are concerned, they have a higher power output but lower efficiency than a similar steam injection cycle. For evaporative cycles, water can evaporate below its boiling point; thus, it does not require extensive heating (i.e., superheaters). The compressed air counter-contacts the heated water surface by placing the evaporator after the compressor. This arrangement causes molecules to collide, allowing water molecules to escape and overcome the vapor pressure [30].

Humidification can also be used for other applications, such as cooling and overcoming compressor sizing limitations, a process known as wet compression [31,32].

CARSOXY [33] is a new set of gas turbine working fluid blends created by combining all of these components, namely carbon dioxide, argon, steam, and oxygen. Previous research [33] has demonstrated that using humidification and evaporation methods can improve the performance of a gas turbine. Furthermore, when a CARSOXY mixture is used as the working fluid, heat recovery methods are used, thus increasing cycle efficiency. High techno-economic benefits can be achieved if a suitable cycle configuration is used under the specific operating condition. Furthermore, further carbon capture and storage techniques, which are considered e imperative requirements for future fossil-fueled gas turbines [34,35], have the potential to improve efficiency while maintaining a relatively low level of CO_2 emissions by using a CO_2–Ar–steam mixture. However, increasing the amount of CO_2 in a gas turbine's working fluid can cause carbon hydrates and blockage in cooling channels in the turbine blades [36], necessitating extra caution in determining the right amount of CO_2 in the working fluid.

As discussed, evaluating the proposed novel approaches of utilizing alternative working fluids in gas turbines is based on the implementation simplicity, cycle efficiency, power outputs, economic sustainability, and acceptable combustibility criteria. Therefore, within those criteria, this paper provides an in-depth examination of the current state of the art in alternative working fluids utilized in the power generation industry (i.e., gas turbines, combustion).

In addition, applying alternative working fluids in power plants implies using separation processes (i.e., air separation, carbon capturing, steam methane reforming, water gas shifting, pressure swing adsorption, distillation, etc.). Therefore, promoting the concept of alternative working fluids to the industry is determined by the feasibility level of implementing those separation processes. Thus, this paper highlights the current contributions of integrating separation techniques, such as air separation, steam–methane reforming, and water–gas shifting to the power generation industry to facilitate a continuous and adequate supply of alternative working fluids.

As far as the integration of separation processes to gas turbines is concerned, there are two major schemes to increase the cycle efficiency or/and reduce the greenhouse effects. The first scheme uses alternative fuels, such as decarbonized fuel blends (i.e., H_2 and NH_3/H_2), amongst many other examples in the literature [37,38]. The second scheme is to use alternative working fluids, which requires the implementation of separation processes to ensure a continuous supply of the components of the alternative working fluid. Therefore, this paper spots the integrated separation techniques to supply the components of the alternative working fluids (oxidized gas turbine cycles, H_2O for humidified gas turbine cycles, CO_2-diluted gas turbine cycles, and inert gas-enhanced gas turbine cycles).

In addition, air may be taken from a gas turbine facility for various reasons, such as for use as a feed to an air separation unit, bleed cooling air for the turbine, or other pressured air requirements inside the facility. The ejected air (Figure 1) includes useful heat recovered by boiling liquid fluids at discrete temperature levels or by sensible heat transfer to other fluids. Solvent regeneration, a process characterized by a gas–liquid absorption stage followed by heat transfer to the liquid to desorb a gaseous product or pollutant, is one type of use for recovered heat. The following unit activities, which may be found in hydrocarbon gasification or hydrocarbon processing facilities, are examples of processes that potentially benefit from this heat integration, including the regeneration of a cryogenic air separation unit's liquid-based air pretreatment system. Extraction air heat recovery might benefit a liquid-based absorption phase for removing impurities from the air supply stream to an ASU. The heated air would be cooled by an absorption column against the liquid bottoms in one embodiment. The cooled air would enter the column and come into contact with the liquid absorbent, absorbing the pollutants in the air stream. The pollutants would be desorbed from the absorbent liquid, then returned to the absorbent column through the air to begin the absorbent heating process. For better efficiency removal, the absorbent system may include one or more fluids in several absorption phases or utilize specialized absorbents to remove specific pollutants in the air stream. Heating from other sources and a pressure decrease in conjunction with heating might be used to desorb contaminants from absorbents. Impurities are desorbed by heating. A gas cleaning solvent, such as that used in the syngas cleanup process, is regenerated. Absorption devices to remove acid gases from the partial oxidation process (POX) plants are widely recognized. Extraction air heat might be utilized in place of or in addition to other heat sources for solvent regeneration. A solvent-based system might be used to recover a salable by-product such as carbon dioxide from a facility's off-gas stream. Again, heat recovered from the extracted air would be used to liberate the product stream from the solvent, either whole or partially.

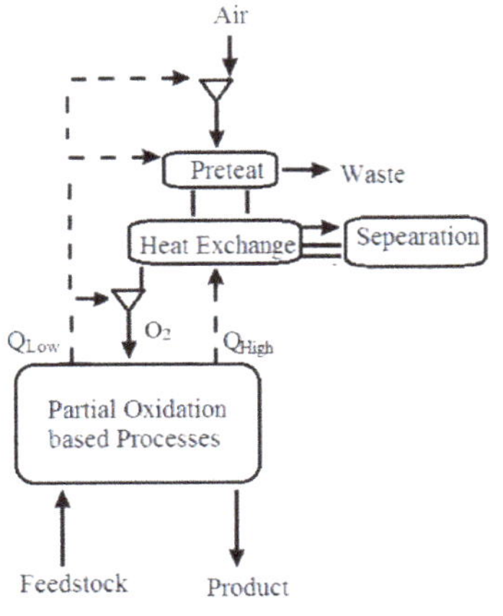

Figure 1. ASU of Chemical PSA [39].

As illustrated in Figure 2, some of these notions can be combined. Heat can be recovered from extracted air by contacting it indirectly with a process fluid or transferring heat from the air to a working fluid such as steam or inert gas. High-level heat from the extracted air source is transferred to a nitrogen stream that returns to the gas turbine in this case. Contact with the rich bottoms of an absorption column used to pretreat air supply to the ASU further cools the extracted air. Other absorption systems in the plant's POX or POX product workup parts might potentially perform this heat transfer phase. The high-level heat recovery phase may be omitted, and all of the extracted air heat may be utilized for absorber renewal, depending on the solvent and substance to be absorbed. Suppose the pressure was considerably different from the supplement air source. In that case, the extracted air might be mixed with supplementary compressed air entering the pretreatment process, or it may be handled independently. Carbon dioxide is collected as a by-product from the air pretreatment region in this embodiment. The carbon dioxide might be treated as a by-product for sale or utilized within the plant. Returning the carbon dioxide to the gas turbine as a diluent is one example.

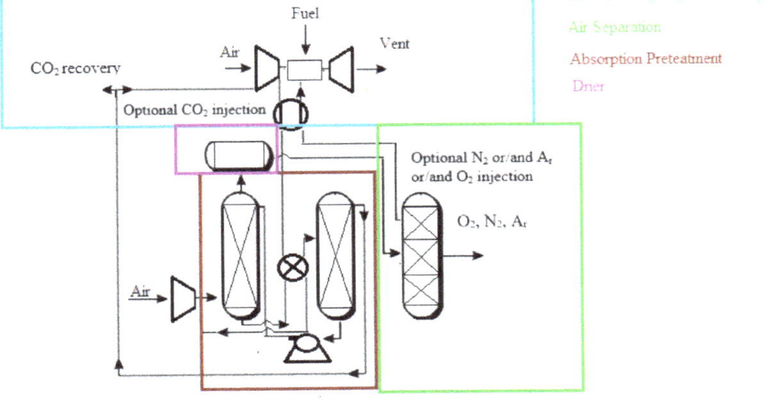

Figure 2. Application of extracted air for heat recovery [39].

2. Carbon Capture and Storage (CCS)

The most prevalent carbon dioxide capturing techniques are discussed in this section. Pre-combustion, post-combustion, and oxyfuel combustion are the three main types. Based on multidisciplinary aspects, the literature has evaluated these techniques (technology maturity state, economic sustainability, advantages, and disadvantages).

2.1. The Main Categories of the CCS

CO_2 is captured from flue gases in post-combustion CCS with no significant changes required. Post-combustion CCS can be easily integrated into existing power plants [40]; however, the electricity costs are increased by approximately 70% [41].

The Technology Centre Mongstad in Norway [42] is one of the most prominent examples of deploying post-combustion CCS commercially. According to the reference [42], the process begins by separating liquid and solid particulates from combustion flue gases in a separation unit, as shown in Figure 3. Flue gases rise from the absorber's bottom against a counter-current stream of lean solution. This allows CO_2 to be absorbed and the treated flue gases to exit the absorber at the top. Meanwhile, the CO_2-rich solution travels from the absorber's bottom to the stripper's top. Against a counter-current water vapor stream, the CO_2-rich solution flows downward. The water vapor stream absorbs the majority of the remaining CO_2. Finally, CO_2 is separated from water vapor using a condensation process.

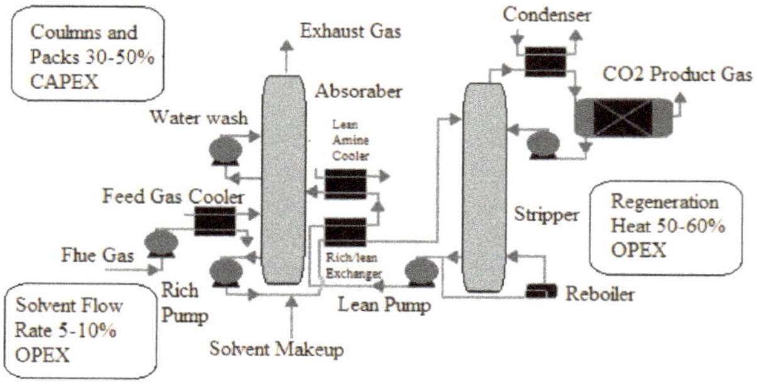

Figure 3. A simplified diagram of a reactive solvent-based post-combustion capture process [42].

Pre-combustion CCS, unlike post-combustion CCS, is difficult to integrate into existing power plants. It necessitates extensive pretreatment, particularly for coal-fired power plants [43], which reduces the system's compactness. On the other hand, the efficiency losses are significantly lower than those seen in post-combustion [44]. The pre-combustion CCS process begins with a coal gasification process, in which coal is converted into hydrogen, carbon monoxide, and carbon dioxide, as shown in Figure 4. This is followed by a water–gas shift reaction, which converts carbon monoxide to carbon dioxide and increases hydrogen production. Carbon dioxide can be captured at this stage using absorption, adsorption, membrane separation, hydrate-based separation, and cryogenic distillation techniques. Natural gas fuel is processed in the same way. The gasification process, on the other hand, is replaced by a reforming process. The CO_2 removal efficiency of natural gas fuel pre-combustion CCS can reach up to 80% [35].

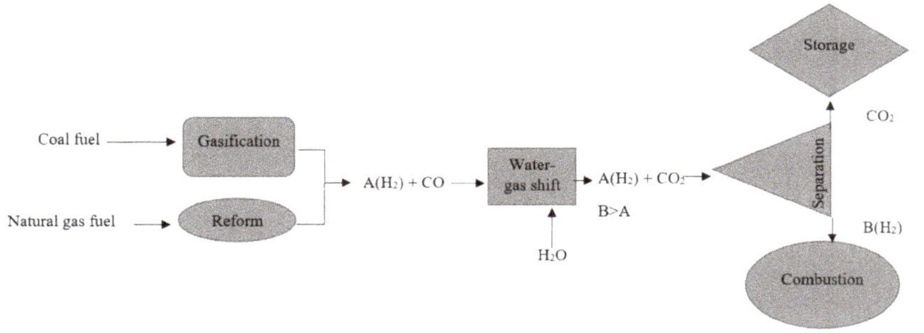

Figure 4. CCS of pre-combustion [44].

Regarding oxyfuel combustion in CCS, the air is not directly supplied to the combustion chamber, as seen in Figure 5. It is first separated into argon, oxygen, nitrogen, and other gases in an air separation unit. The oxygen is extracted and fed into a combustion chamber. Carbon dioxide is highly concentrated in oxyfuel combustion flue gases [41], thus allowing easier carbon capturing. However, the use of the ASU for an oxyfuel lignite-fired power plant, for example, has been reported to reduce plant efficiency by up to 10% [44].

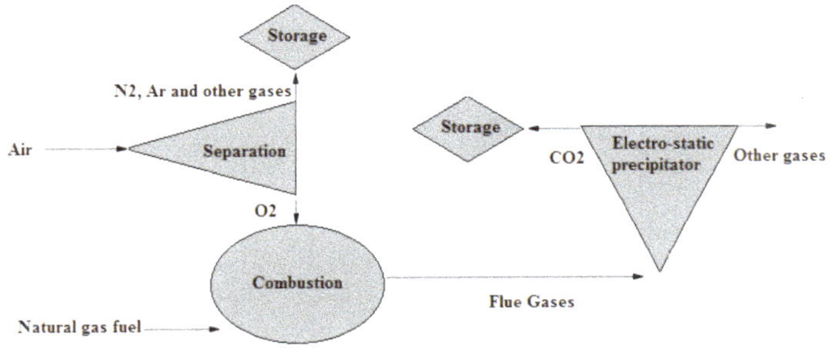

Figure 5. CCS of oxyfuel combustion [44].

Table 1 summarizes the three CCS technologies' maturity states, economic status, and benefits and drawbacks [44].

2.2. Methods for Separating Carbon Dioxide

CO_2 is separated by absorption using liquid sorbents, and stripping techniques can be used to extract the latter. Monoethanolamine (MEA) ensures a CO_2 absorption efficiency of 90% [45]. Adsorption [45] is the technique used when the sorbent is in the solid phase (such as hydrotalcite, molecular sieves, and so on). Another well-known technique is membrane separation, which allows CO_2 to pass through a porous composite polymer selectively. Because different gases have different molecule sizes, pores are carefully sized to match the size of the CO_2 molecule [46]. Table 2 [46] compares and contrasts the benefits and drawbacks of the most common CO_2 separation techniques.

Table 1. The current state of CCS technologies.

Capturing Technology	Technology Status	Cost	CO_2 Concentration	Thermal Efficiency	Pros	Cons
Post-combustion	Fully Developed	The lowest	The lowest	The highest	Easily integrated into existing power plants.	The electricity costs are increased by approximately 70%
Pre-combustion	Fully Developed	Medium	Medium	The lowest	The efficiency losses are significantly lower than those seen in post-combustion. Enables hydrogen production. The CO_2 removal efficiency of natural gas fuel pre-combustion CCS can reach up to 80%	Difficult to integrate into existing power plants. Reduces the system's compactness.
Oxyfuel combustion	Developed air separation	The highest	The highest	Medium	Eliminates NO_x emission of the power plant.	Reduces plant efficiency by up to 10%

Table 2. CO_2 pros and cons.

CO_2 Separation Method	Pros	Cons
Absorption	• The most developed technology • The highest CO_2 recovery efficiency (approximately 90%)	• Low efficiency for low CO_2 concentration, i.e., Post-combustion • High heat requirements
Adsorption	• Recyclable absorbent • Relatively high CO_2 recovery efficiency (approximately 85%)	• High heat requirements
Membrane separation	• Relatively high CO_2 recovery efficiency (approximately 80%)	• Operational issues
Hydrate-bases separation	• Negligible energy penalty	• Not fully developed technology
Cryogenic distillation	• Fully developed technology	• High-energy penalty

2.3. CO_2 Storage

Structural and stratigraphic, residual, solubility, and mineral trapping are the four primary trapping mechanisms for CO_2 storage [47]. Once CO_2 is injected underground, it rises to the top of geological structures, owing to the buoyancy effect. Still, it remains below the impermeable caprock with structural and stratigraphic trapping, which is the most prominent trapping process. The injected CO_2 displaces formation fluid as it passes through the formation rock in residual trapping. Due to the capillary force, the displaced fluid disconnects and traps the remaining CO_2 inside the pores of rocks [48]. The saturation of trapped CO_2 in the residual trapping mechanism is at least 10% and can reach more than 30% of the pore volume in some reservoir rocks [49]. CO_2 dissolves in formation fluids and becomes immobile, resulting in a reduction in the amount of free CO_2 [36]. The density of the formation fluid will be slightly increased by about 1% due to the dissolved CO_2. Such a small density differential [36] is adequate to induce convection flow, which is also beneficial to CO_2 trapping. CO_2 solubility in groundwater varies between 2% and 6%, decreasing with increasing temperature and salinity [50]. CO_2 is held in the mineral trapping mechanism by geochemical processes in the reservoir, which generally precipitate as carbonate, effectively trapping CO_2 in immobile secondary phases [50].

In a period ranging from 1 to 10,000 years, various trapping processes play distinct roles in CO_2 storage. Structural trapping is vital in the early stages of CO_2 storage, but its impact fades over time. Within decades, residual trapping and solubility trapping significantly influence and lock up a large quantity of CO_2 for thousands of years. Mineral trapping begins to have a substantial influence around a hundred years and plays a critical function in a geological timeframe.

For various reasons, storing CO_2 in hydrocarbon reservoirs is one of the most preferred CO_2 storage options. For example, oil and gas reservoirs contain a large quantity of existing equipment on the surface and underground that may be utilized for CO_2 storage with only minimal modifications. Furthermore, the caprock's seal quality and integrity are assured, having been thoroughly defined during the exploration and production process. Moreover, because of the long-term extraction of oil and gas, the magnitude of pressure perturbations and associated stress shifts in depleted oil and gas reservoirs is significantly smaller than in aquifers. Depleted gas reservoirs are more suitable for CCS than depleted oil reservoirs because of gas's higher ultimate recovery and compressibility and a bigger storage capacity per pore volume. When comparing the types of reservoirs utilized in this type of storage, condensate gas reservoirs outperform wet and dry gas reservoirs due to the small amount of residual gas, the phase behavior of the combination of condensate gas, and CO_2, and its high injectivity. Additionally, the amount of CO_2 sequestered per pore volume in depleted condensate reservoirs is relatively high: around 13 times that of a comparable aquifer. Yet, the phase transition may occur in depleted condensate reservoirs but not in dry and wet gas reservoirs; therefore, caution is advised.

3. Air Separation Unit (ASU)

The concept of oxyfuel combustion is based on utilizing pure oxygen as the primary working fluid while eliminating nitrogen to avoid NO_x emissions. As previously mentioned, pure oxygen is typically accompanied by inert gases, such as argon, for its high heat capacity, to increase the cycle efficiency. Therefore, to separate oxygen and argon from the air while excluding nitrogen from the combustion process (i.e., to avoid NO_x emission), air separation techniques are integrated into oxyfuel combustion-based power plants [51].

3.1. Types of Cryogenic Distillation

Cryogenic distillation can be performed using a single distillation column or multiple distillation columns. The multi-column process [52] is, however, the traditional method. A low-pressure column, a high-pressure column, and a side rectifier or strippers are the main components of this process. The cost of multi-column distillation remains a challenge.

However, several cost-effective improvements have been proposed in the literature [52] and are detailed in the following sections.

3.1.1. Thermally Linked Distillation Column, with a Side Rectifier

According to the reference [53], A ternary mixture made up of A, B, and C components (i.e., N_2, Ar, and O_2 in air, respectively) are separated, as shown in Figure 6. The most volatile component is A (i.e., N_2 in the air), followed by B and C (i.e., Ar and O_2 in air, respectively), the intermediate and least volatile components, respectively. The mixture is supplied to the main distillation column, which extracts component A (i.e., N_2 in air) from the top. From the bottom, component C (i.e., O_2 in air) is extracted. The ABC (i.e., air) mixture's main feed is located approximately in the middle of the distillation column. The side feed is connected to the bottom of the rectifier and located just beneath the main feed. B and C (i.e., Ar and O_2 in air, respectively) are the components of the side feed mixture. B (i.e., Ar in the air) is produced from the top of the side rectifier, while the C (i.e., O_2 in the air) component accumulates at the bottom of the side rectifier. At the same location as the side feed, component C (i.e., O_2 in the air) from the side rectifier is recycled back into the main column. When the relative volatility of the two components (B and C (i.e., Ar and O_2 in the air)) in the side feed is low, this method is used.

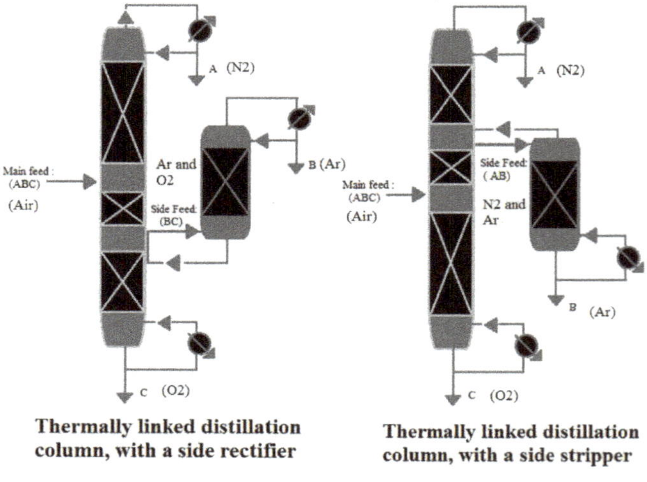

Figure 6. Thermally linked distillation column, reproduced from the reference [53].

3.1.2. Thermally Linked Distillation Column, with a Side Stripper

A side stripper is a better choice if the low relative volatility is between (A and B) rather than (C and B). The side feed is linked to the top of the stripper and is located above the main feed, as shown in Figure 6. The side feed mixture comprises A and B, with B produced at the stripper's bottom, whereas A is returned to the main column.

3.1.3. Side Rectifier vs. Side Stripper

The concept of using ternary mixture distillation on air is depicted in Table 3. Because argon and oxygen (B and C) have lower relative volatility than argon and nitrogen (B and A) [54], a side rectifier is a more proper choice for air separation.

3.2. Low-Pressure (LP) Distillation Column vs. Elevated-Pressure (EP) Distillation Column

Low-pressure distillation columns are used if the by-products of nitrogen production (oxygen and argon) are not stored and discharged near atmospheric pressure. The storage of oxygen and argon, on the other hand, necessitates an additional increase in pressure. As a result, evaluated-pressure distillation columns are utilized. To put it another way, if the by-products of producing nitrogen are compressed and used as main products, an elevated-pressure distillation column is required [52]. The elevated-pressure distillation column provides better matching between the ASU and the gas turbine unit; for example, the working fluid of CARSOXY gas turbines is required to operate at relatively high pressure; thus, elevated-pressure (EP) distillation columns are required [50].

Table 3. Air components relative volatility [54].

Component	Volatility	Boiling Point	in Air	Relative Volatility
A	High	Low (−195.8 °C)	Nitrogen	The relative volatility of argon and oxygen (B and C) is lower than that of argon and nitrogen (B and A).
B	Intermediate	Intermediate (−185.8 °C)	Argon	
C	Low	High (−183 °C)	Oxygen	

3.3. Single Distillation Column vs. Double Distillation Columns

In comparison to conventional double distillation columns, self-heat recuperation of a single distillation column saves 36% of energy consumption, according to the reference [46]. Liquefaction and air compression are the significant losses in a double-column air separation process. Double distillation columns, on the other hand, are still the most popular. Table 4 provides a critical assessment of the air separation methods.

Table 4. Pros and cons of the air separation methods (distillation).

Air Separation Method.	Pros	Cons
Cryogenic distillation	Can be performed using a single distillation column or multiple distillation columns, depending on the specifications and requirements of the integrated power plant	The cost of multi-column distillation remains a challenge
Thermally linked distillation column, with a side rectifier	Used for alternative working fluids with high contents of oxygen and argon.	Excessive production of nitrogen for nitrogen-independent working fluid.
Thermally linked distillation column, with a side stripper	–	Because argon and oxygen (B and C) have a lower relative volatility than argon and nitrogen (B and A) [54], a side rectifier is a more proper choice for air separation.
Low-pressure (LP) distillation	Less energy consumption compared to the EP distillation	Not suitable for gas turbines operating at high compression ratios
The elevated-pressure distillation column	Provides better matching between the ASU and the gas turbine unit compared to Low pressure (LP) distillation	High energy consumption
Single distillation column	In comparison to conventional double distillation columns, self-heat recuperation of a single distillation column saves 36% of energy consumption	Less popular, compared to the double distillation columns
Double distillation columns	More popular compared to the single distillation column	Liquefaction and air compression cause significant losses

3.3.1. Conventional Argon Production

Air is fed to the ASU through a heat exchanger at about 5.5–6.5 bar, as shown in Figure 7, to be cooled by its products. The first stage is a high-pressure (HP) stage, and the second stage is a low-pressure (LP) stage in a two-stage distillation column that separates oxygen and nitrogen. At the HP stage, the air feed is separated into N_2 and oxygen-enriched liquid (liquid oxygen (LOX)) at the HP stage. The latter is passed through the LP stage, where it is separated into N_2 and O_2. There is a temperature difference between the two stages due to the pressure difference. This allows heat exchange between the two stages (between LP and HP), in which the N_2 vapor-boiling stream at the top of the HP stage is then condensed by the colder liquid O_2 at the bottom of the LP stage. From the top of the HP column, N_2 is produced. The highest argon concentration is found in the lower section of the LP column due to its intermediate boiling temperature. A vapor stream is drawn to feed the rectifier from this location. The vapor argon-boiling stream at the top of the rectifier is condensed by the colder LOX liquid, similar to the main distillation column. LOX that has been vaporized is returned to the LP column.

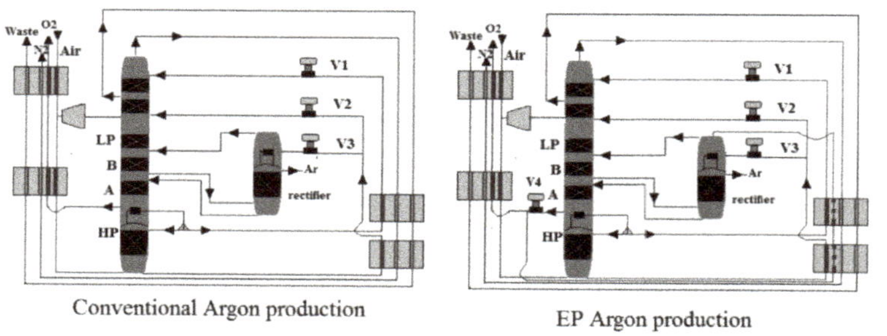

Figure 7. Thermally linked distillation column, reproduced from the reference [53].

3.3.2. EP Argon Production

As shown in Figure 7, oxygen in the liquid phase is drawn from the bottom of the LP to provide total condensing duty, while argon is drawn from the top of the rectifier column to improve argon recovery due to the elevated-pressure process's tight volatility. The vaporized oxygen serves a second purpose after completing the total condensing duty. As the EP process reaches higher temperatures, it is used as a coolant agent in the system. Because of the heat pump effect between the argon rectifier and the oxygen at the bottom of the LP column, the pressure at the top of the rectifier remains lower than the pressure at the bottom of the LP column, requiring no compressor.

3.4. Membrane Air Separation

One of the most important ways to generate oxygen from air separation is using new ceramic membranes with mixed electronic ionic conducting (MIEC) properties. Figure 8 depicts a schematic of an oxyfuel combustion power plant using an MIEC membrane. As shown in Figure 8, an air compressor may pressurize air to a pressure of 10 bar. Both the membrane and the air must be subjected to temperatures exceeding 800 °C, as well as an oxygen partial pressure ratio across the membrane, to accomplish air separation. A combination of air compression, flue gas recycling, and boiler heat exchange can provide the heat necessary to reach this high temperature. Surprisingly, the front end of an oxyfuel combustion plant can accommodate the ceramic membrane module. Because of the continuous oxygen partial pressure drop across the membrane, which is one of the technological aspects of such integration, using recovered CO_2 from the combustion process might result in a strong oxygen driving force. Oxygen penetrates selectively through a membrane into the recycled CO_2 stream from the compressed air side, enriching it to around 20% (*v/v*) [55].

After that, the oxygen-enriched CO_2 stream may be burned directly in the boiler, resulting in superheated steam that can be used to power steam turbines and generate electricity.

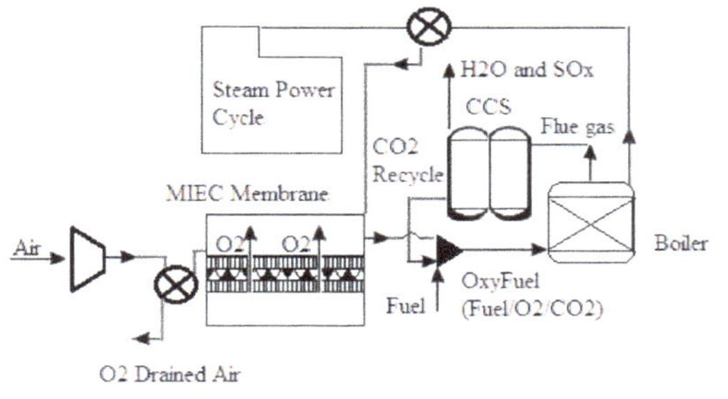

Figure 8. Diagram of an oxyfuel power plant using an MIEC membrane module [55].

As shown in Figure 9, there are two distinct approaches for integrating an MIEC membrane into the oxyfuel combustion process. The notion of enabling recirculated flue gas to pass over the membrane surface is used in Figure 9a, allowing oxygen to be absorbed directly. This so-called four-end approach [56–58] is intended for Integrated Gasification Combined Cycle (IGCC) processes; however, CO_2-resistant membrane materials such as Ba0.5Sr0.5CO0.8Fe0.2O3I (BSCF) do not exist. As a result, BSCF membranes can only be used when the membrane material is not in direct contact with the flue gas. On the other hand, the three-end idea [59,60] depicted in Figure 9b appears to be an appealing alternative, as it avoids direct contact between the membrane and the flue gas. A vacuum pump is used to remove oxygen from the membrane module in this scenario. Although the first concept achieves a higher thermal efficiency in the oxyfuel combustion process, the three-end concept is more likely to be feasible because no membrane materials exist that can withstand direct contact with the flue gas; therefore, membrane material development is critical [61–63]. Kneer et al. [56] effectively provided the requirements of coal combustion in a CO_2/O_2 environment, including associated burner design, as well as the cleaning of hot flue gas from oxycoal combustion, in their article on membrane-based air separation modules and their design for oxycoal circumstances.

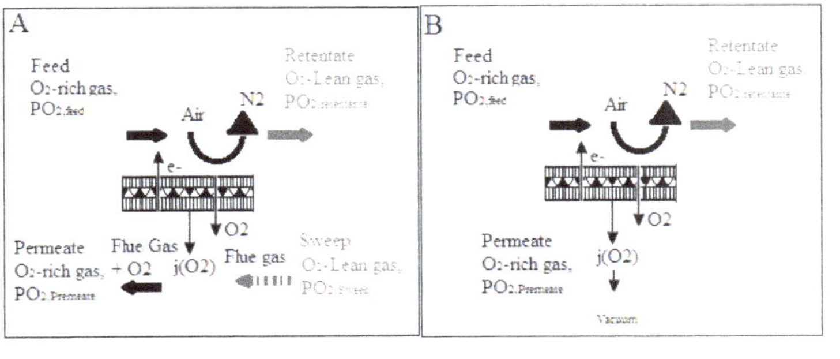

Figure 9. (**A**) four-end approach and (**B**) the three-end concept for oxygen separation in oxyfuel processes [62].

It is worth noting that there are several problems with using this innovative approach for obtaining oxygen from air separation. To offer high oxygen penetration flux, the membrane separation system must first be run at temperatures exceeding 900 °C. As a result, because the heated air includes around 80% (*v/v*) nitrogen as waste gas, there will be significant energy penalties in this operation. Although process efficiency losses can never be entirely recovered, energy losses can be minimized by adequately implementing heat recovery. Second, while the strong oxygen driving force of high CO_2 partial pressure is appealing to membranes, the aggressive circumstances would lead these membrane materials to break down, limiting their efficacy in generating oxygen for the oxyfuel combustion process. The high need for oxygen for the oxyfuel combustion process is likely to lead to the deployment of oxygen membrane separation technology in the near future; nevertheless, continuing research and development is critical to overcome the significant challenges. In summary, the oxyfuel power plant is an appealing choice due to a considerable cost decrease in oxygen generation when carbon capture becomes a requirement in the future. As a result, ceramic membranes are thought to be an effective high-temperature air separation technology that might result in a significant cost decrease in the generation of oxygen for this application.

It should be mentioned that an IGCC coal-fired power station combines two technologies, coal gasification and combined cycle, to produce energy in the most efficient manner possible [63,64]. Traditionally, the oxygen supply for coal gasification has been obtained by cryogenic distillation; however, it has been observed that this technique has many inefficiencies [65]. The use of solid oxygen carriers as an alternate way of producing oxygen has also been investigated, including $CaSO_4$ [66,67] and other oxygen carriers such as manganese [68], nickel [69], and iron [70,71] oxides. However, there are drawbacks to using this technique, including sluggish response speeds, solid handling, and a huge solid oxygen carrier inventory [70,71]. Because these membrane modules have no moving parts and are thus simple to run and maintain, introducing a revolutionary technology based on ceramic membranes for oxygen generation from air separation is thought to enhance the efficiency of IGCC power plants.

Table 5 provides a critical assessment of the membrane air separation methods.

Table 5. Pros and cons of the membrane air separation methods.

Membrane Air Separation Method	Pros	Cons
MIEC membrane	• Can be well-integrated with CCS units within the power plant. • Because these membrane modules have no moving parts, they are simple to run and maintain.	• High temperature and pressure are required (approximately 800 °C and 10 bar). • The aggressive circum-stances would lead these membrane materials to break down, limiting their efficacy in generating ox-ygen for the oxyfuel com-bustion process.
Four-end approach	• Suitable for Integrated Gasification Combined Cycle (IGCC) processes.	• Can only be used when the membrane material is not in direct contact with the flue gas.
Three-end approach	• Avoids direct contact between the membrane and the flue gas.	• A vacuum pump is required to remove oxygen from the membrane.

4. Steam Methane Reforming (SMR)

SMR is an essential step in the carbon capture and storage (CCS) process, as discussed in Section 5.1. Furthermore, it is one of the fully developed technologies for producing hydrogen. The results and conditions in real life are identical to those in theory [72]. When a light hydrocarbon fuel (such as methane) reacts with steam, it produces hydrogen as the primary product and carbon monoxide and carbon dioxide as by-products. Since WGS processes accompany SMR in most applications, some literature includes WGS as a step in SMR [72]. Two endothermic reactions occur during the reforming process. One mole of methane to one mole of steam requires 206 kJ in the first reaction. This yields 3 moles of hydrogen and 1 mole of carbon monoxide as a by-product. WGS makes use of the hydrogen produced by this reaction by converting carbon monoxide to hydrogen and carbon dioxide as a by-product [73]. The second reaction, which produces hydrogen and carbon dioxide directly instead of one mole of methane reacting with two moles of steam, does not require WGS because it directly produces hydrogen and carbon dioxide. It uses 165 kJ and emits 4 moles of hydrogen and 1 mole of CO_2. Finally, the CCS process is used to capture all of the carbon dioxide produced during the entire process. Figure 10 depicts the whole procedure.

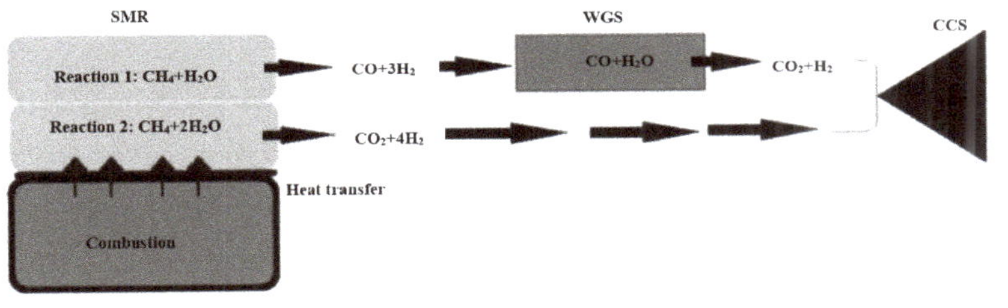

Figure 10. SMR process, reproduced from the reference [74].

The SMR process is characterized by harsh reaction conditions [74]. Because reactions 1 and 2 are both endothermic, the temperature rises to around 1000 K. It has even been described as a "brutal" process [73]. Both reactions take place in a catalytic environment. Table 6 [74] explains that the catalyst must be carefully selected to withstand these extreme conditions.

Table 6. SMR catalyst [75].

Category	Temperature Range (°C)
nickel-based	550–850
carbon-based	300–900
zinc-based	300–500
aluminum-based	750–800
cerium-based	300–800
zirconium-based	250–500

Because of well-developed methane infrastructures and methane's favorable hydrogen-to-carbon ratio, methane steam reforming is commonly regarded as a readily available technique for producing hydrogen on a large scale [76]. Fuel cell technology needs small, low-cost reformers [77,78]. Compact reformers should be operated at low temperatures (<1000 K) and pressures (3 bar). As a result, the present large-scale reformer technology, which operates at high temperatures (>1100 K) and pressures, is incompatible with smaller-scale reformers for fuel cell applications [79]. Methane steam reforming is a substantially endothermic process. The exothermic water–gas shift process is also included, which is more favorable at low temperatures (500–850 K). The benefit of executing the steam reforming process at low operating temperatures is that the applied temperature promotes the water–gas shift reaction by suppressing the CO quantity [80]. Creating an active catalyst capable of achieving conversion up to equilibrium levels is the primary disadvantage of methane steam reforming processes at low temperatures [81–83]. Several groups investigated catalyst performance in methane steam reforming at temperatures above 1000 K [84–86].

5. Water–Gas Shifting (WGS)

Water–gas shifting is a chemical reaction that converts carbon monoxide and water into carbon dioxide and hydrogen. It is one of the most essential processes in carbon capture and storage. About 40 kJ is generated when one mole of carbon monoxide is reacted in the WGS reaction. Two types of catalysts (iron-based and copper-based) are commonly used to speed up the reaction [87].

5.1. WGS Applications

Water–gas shifting is an area of interest whenever carbon monoxide is an unwanted by-product gas. It is, in fact, a very reliable method of purifying hydrogen produced by steam hydrocarbon reforming processes. This process is typically integrated with the SMR process, Reactions (2) and (3) [26], as discussed in Section 7.

$$\text{Steam reforming of methane: } H_2O + CH_4 \leftrightarrow CO + 3H_2 \quad (2)$$

$$\text{Water-gas shift reaction } H_2O + CO \leftrightarrow CO_2 + H_2 \quad (3)$$

Ammonia production is another industrial application that relies heavily on water–gas shifting to prevent carbon monoxide from reacting with the catalyst [88]. Ammonia production is likely to be a part of the gas turbine cycle, primarily if evaporation injection techniques are used [88].

5.2. WGS Catalyst

The two most important factors to consider when selecting catalysts for the WGS process are sulfur tolerance and operation temperature. Iron-based catalysts have a low sulfur tolerance and operate at a high temperature (approximately 680 K). Carbon-based catalysts are used at lower operating temperatures (around 480 K). On the other hand, it has a low tolerance level for sulfur [89]. The main properties of the iron-based and copper-based catalysts are shown in Table 7 [89].

Table 7. WGS catalysts.

Property	Composition	Stabilizer	Promoter
Iron-based catalyst	74.2% Fe_2O_3, 10% Cr_2O_3, 0.2% MgO, and the remaining is to balance volatiles	Cr_2O_3	—
Copper-based catalyst	24% ZnO, 24% Cr_2O_3, 3% Mn, Al, and Mg-oxides. The remaining is to balance volatiles	Cr_2O_3	ZnO

5.3. WGS Reactors

The conventional WGS process consists of two reactors. It uses an iron-based catalyst in a high-temperature reactor (HTWGS). The high-temperature reactor's products are cooled before being purified from CO in a low-temperature reactor (LTWGS) with a copper-based catalyst. This option is ideal if the reactor's inlet boundary condition is high, and the desired product (H_2 and CO_2) must be produced at a low temperature with very low CO residuals.

Figure 11 [51] shows a WGS reactor with a shell-tube design. A hydrogen-selective membrane tube is inserted into a shell. The catalytic WGS reaction is carried out around the tube (inside the shell). The membrane performs the separation function by allowing only H_2 to pass through (inside the tube), while the remaining WGS products are extracted as a CO_2-rich mixture outside the tube.

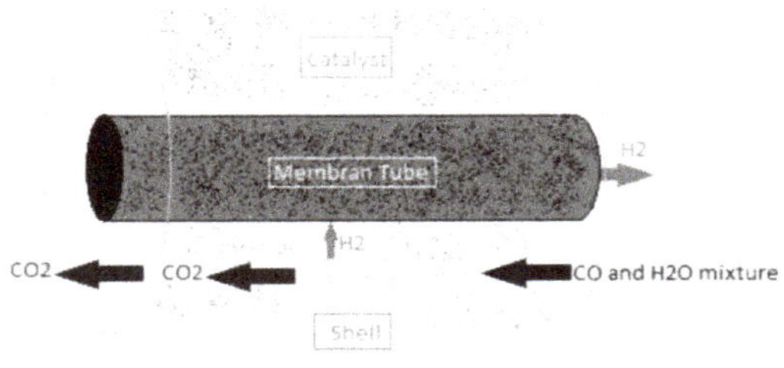

Figure 11. Shell-tube WGS reactor, reproduced from the reference [80].

6. Helium as a Working Fluid

Many nations, including Russia, Europe, South Africa, Japan, and the United States, are studying direct and indirect Closed Brayton Cycle (CBC)-based high-temperature gas-cooled reactors (HTGRs) that employ helium as a coolant for energy conversion. Because of its outstanding transport characteristics, helium is regarded as the best coolant. According to the data, helium has been utilized as the working fluid and coolant in 9 of the 15 closed Brayton cycle nuclear reactor systems built across the world since the 1960s [90]. However, because helium is less compressible than air, more stages are necessary to compress it to the desired pressure ratio in an axial compressor. As a result, every helium compressor ever built has a large number of stages. The Oberhausen II type 50 MW reactor's high- and low-pressure compressors each contain 25 stages [90]. As a result, the helium turbo equipment is rather substantial in terms of both size and weight. This situation is unfavorable because it raises various losses, such as the formation of a multi-stage narrow fluid flow channel. As a result, there is an increase in aerodynamic loss due to leakage, separation, and boundary layer loss. As a result, the stator and rotor stages are mismatched. Aside from that, compressors with narrow rotors cause dynamic difficulties [91,92]. As a result, there is a pressing need to address this problem since smaller shafts with two spools turbo machinery enhance plant efficiency over longer shafts because they are easier to maintain, perform better dynamically, and have high rigidity [93]. An extensive study is being carried out to identify the most viable working fluid for utilization in HTGR systems with increased thermal efficiency, electrical productivity, decreased size, and lowest rotational loss to solve the problem mentioned above.

On the other hand, researchers devised a novel design approach for a heavily loaded helium compressor to minimize the number of stages in the helium compressor. This design approach is based on changing the form of the airfoil or blade when torsional velocity, axial velocity, and negative pre-spin increase. As a result, stage loading increases [15]. Numerical simulations of a heavily loaded helium compressor were carried out using the commercial program Ansys, and it was discovered that this design method can minimize the number of stages. Still, no experiment has been carried out using it. Noble gases and their mixes are not yet used as a working fluid in CBCs. Certain factors influence the choice of working fluid for the HTGR system's power conversion unit. Noble gases and their mixes are the subjects of many studies for use as a working fluid in CBC HTGR plants and axial flow turbomachinery [94,95]. Many investigations have been carried out to evaluate the heat transmission process of the HeXe mixture in triangular channels [95], cylindrical channels [96], beginning pipe sections [97], and heated channels with varied cross-sectional forms [98]. The HeXe gas mixture's thermodynamic characteristics and transport constants were also calculated [98]. HeXe was also used in the Prometheus project. It was based on a Brayton energy conversion loop with a single reactor heat source [99].

7. Conclusions

Local and regional regulations have been established to limit NO_x and CO_2 emissions as part of this global commitment. Most systems responsible for these undesirable emissions are directly linked to power generation, with gas turbines playing a pivotal role. As a result, stationary gas turbines have been widely used in power plants and will continue to be used to meet the growing electricity demand. As a result, novel concepts are needed to reduce emissions while allowing large amounts of power to be generated from these systems. One promising technology for reducing harmful emissions while recirculating CO_2 in the combustion process is alternative working fluids. The idea behind using alternative combustion working fluids is to replace air with oxygen, carbon dioxide, steam, or inert gases, either individually or collectively.

Using carbon dioxide as a working fluid in an oxyfuel gas turbine keeps the temperature of the turbine within acceptable limits and reduces heat losses, all while removing NO_x emissions. Chemical kinetics, flammability regions, and flame compactness, on the other hand, are negatively affected. Increased oxygen levels could lessen the severity of those negative consequences. However, air separation units will be required to provide more oxygen, resulting in additional costs. As a result, it can be concluded that using CO_2- O_2 as a working fluid (without any additional components to improve overall thermodynamic properties) poses significant technical challenges and is unlikely to be implemented on a large scale.

When compared to conventional air-driven simple cycle gas turbines, the main benefit of injecting steam into gas turbines is an increase in cycle efficiency and specific power outputs. However, based on the research on O_2-H_2O-CH_4, it can be concluded that O_2-H_2O-CH_4 power plants do not always achieve higher efficiencies and power outputs than air-driven cycles (i.e., aero-derivative and industrial combined cycles achieve higher efficiencies and power outputs than steam-injected or (O_2-H_2O-CH_4) gas turbines). As a result, humidification techniques should be used with caution to achieve higher efficiency than air-driven cycles. O_2-H_2O-CH_4 combustion/power plants showed more advanced properties, as shown in Table 7. However, those properties (laminar flame speed, CO emissions, and exergy efficiency) are only compared to the literature's O_2-CO_2-CH_4 combustion/power plants. They must be benchmarked against air-CH_4 combustion/power plants in future studies.

Due to their higher specific heat ratios and potential for increasing cycle efficiency compared to air, argon, xenon, and helium have been studied as working fluids in the context of using inert gases in combustion and power generation. In comparison to air-driven cycles, using helium as a working fluid in closed cycles has the potential to improve cycle efficiency. However, these techniques have not yet matured to the point where they can be used on a large industrial scale (i.e., helium leakage is not easily controlled due to its low molecular weight).

Due to the high specific heat ratio of xenon (1.677) and argon (1.667) compared to nitrogen (1.401), using oxygen–argon and oxygen–xenon mixtures is expected to increase cycle efficiency while eliminating NO_x. However, because argon has a higher abundance and mixing rate than xenon, it is more likely to be used in combustion in the future.

The advantages of using carbon dioxide to keep turbine temperatures within acceptable ranges, argon to increase specific heat ratio, steam to increase mass flow rate, and oxygen to eliminate NO_x emissions are incorporated into the CARSOXY working fluid concept.

As discussed, evaluating the proposed novel approaches of utilizing alternative working fluids in gas turbines is based on the implementation simplicity, cycle efficiency, power outputs, economic sustainability, and acceptable combustibility criteria. Therefore, within those criteria, this paper provides an in-depth examination of the current state of the art in alternative working fluids utilized in the power generation industry (i.e., gas turbines, combustion).

In addition, applying alternative working fluids in power plants implies using separation processes (i.e., air separation, carbon capturing, steam methane reforming, water gas shifting, pressure swing adsorption, distillation, etc.). Therefore, promoting the concept of alternative working fluids to the industry is determined by the feasibility level of implementing those separation processes. Consequently, this paper has highlighted the current contributions of integrating separation techniques, such as air separation, steam methane reforming, and water–gas shifting to the power generation industry to facilitate a continuous and adequate supply of alternative working fluids.

Finally, the decision of choosing the composition of the optimal working fluids can be made through numerical analyses that correlate the individual effect of each component to the overall composition with respect to the gas turbine cycle efficiency. As an example of this practice, the reference [48] has related the components of the alternative working fluid with respect to the cycle efficiency through a three-dimensional efficiency surface that correlates three intervals of variable molar fractions of carbon dioxide, argon, and steam. The highest cycle efficiency is visualized as the highest peak on the efficiency surface within the tested intervals of molar fractions. As a result, the optimal blend was determined by selecting the corresponding molar fractions (to the highest peak).

As highlighted previously, the idea behind using alternative combustion working fluids is to replace air with oxygen, carbon dioxide, steam, or inert gases. Therefore, ASU is utilized to produce the required oxygen and argon content. Steam and carbon dioxide are supplied to the alternative working fluid mixture by integrating humidification and CCS techniques. However, to produce the required molar fractions to achieve the optimal working fluid blends, the operating conditions of the integrated facilities (i.e., ASU, humidification, and CCS facilities) have to be calibrated through performing extensive sensitivity analyses. An example of this practice can be found in the reference [48].

Author Contributions: Conceptualization, O.F.A., A.I.A., M.H.E.-N., M.H., Y.A.O., W.F., A.S.A.-t. and A.V.M.; methodology, O.F.A., A.I.A., M.H.E.-N., M.H., Y.A.O., W.F., A.S.A.-t. and A.V.M.; software, O.F.A., A.I.A., M.H.E.-N., M.H., Y.A.O., W.F., A.S.A.-t. and A.V.M.; validation, O.F.A., A.I.A., M.H.E.-N., M.H., Y.A.O., W.F., A.S.A.-t. and A.V.M.; formal analysis, O.F.A., A.I.A., M.H.E.-N., M.H., Y.A.O., W.F., A.S.A.-t. and A.V.M.; investigation, O.F.A., A.I.A., M.H.E.-N., M.H., Y.A.O., W.F., A.S.A.-t. and A.V.M.; resources, O.F.A., A.I.A., M.H.E.-N., M.H., Y.A.O., W.F., A.S.A.-t. and A.V.M.; data curation, O.F.A., A.I.A., M.H.E.-N., M.H., Y.A.O., W.F., A.S.A.-t. and A.V.M.; writing—original draft preparation, O.F.A., A.I.A., M.H.E.-N., M.H., Y.A.O., W.F., A.S.A.-t. and A.V.M.; writing—review and editing, O.F.A., A.I.A., M.H.E.-N., M.H., Y.A.O., W.F., A.S.A.-t. and A.V.M.; visualization, O.F.A., A.I.A., M.H.E.-N., M.H., Y.A.O., W.F., A.S.A.-t. and A.V.M.; supervision, O.F.A., A.I.A., M.H.E.-N., M.H., Y.A.O., W.F., A.S.A.-t. and A.V.M.; project administration, O.F.A., A.I.A., M.H.E.-N., M.H., Y.A.O., W.F., A.S.A.-t. and A.V.M.; funding acquisition, O.F.A., A.I.A., M.H.E.-N., M.H., Y.A.O., W.F., A.S.A.-t. and A.V.M. All authors have read and agreed to the published version of the manuscript.

Funding: This publication was made possible by NPRP 13 grant # [NPRP13S-0203-200243] from the Qatar National Research Fund (a member of Qatar Foundation). The findings herein reflect the work and are solely the responsibility of the author.

Informed Consent Statement: Not applicable.

Acknowledgments: The authors dedicate this work to Yousif Al-Najjar (May his soul rest in peace), as an acknowledgment of his contribution to enriching the literature with novelties and science.

Conflicts of Interest: The authors declare no conflict of interest.

Abbreviations

CCS	Carbon Capture and Storage
ASU	Air separation units
HTGR	High-temperature gas-cooled nuclear reactors
HR	Heat recovery
CARSOXY	Carbon dioxide, argon, steam, and oxygen
POX	Partial oxidation process
SMR	Steam methane reformer
WGS	Water–gas shift
PSA	Pressure swing adsorption
MEA	Monoethanolamine
LP	Low pressure
EP	Elevated pressure
HP	high pressure
LOX	Liquid oxygen
MEIC	Mixed electronic ionic conducting
IGCC	Integrated gasification combined cycle
CBC	Closed Brayton cycle
NSPS	New source performance standards
BSCF	Ba0.5Sr0.5CO0.8Fe0.2O3I
HTWGS	High-temperature water–gas shift
LTWGS	Low-temperature water–gas shift

References

1. Scheffknecht, G.; Al-Makhadmeh, L.; Schnell, U.; Maier, J. Oxy-fuel coal combustion—A review of the current state-of-the-art. *Int. J. Greenh. Gas Control* **2011**, *5*, S16–S35. [CrossRef]
2. Álvarez, J.F.G.; de Grado, J.G. Study of a modern industrial low pressure turbine for electricity production em-ployed in oxy-combustion cycles with CO_2 capture purposes. *Energy* **2016**, *107*, 734–747. [CrossRef]
3. Tuttle, S.G.; Chaudhuri, S.; Kostka, S., Jr.; Kopp-Vaughan, K.M.; Jensen, T.R.; Cetegen, B.M.; Renfro, M.W. Time-resolved blowoff transition measurements for two-dimensional bluff body-stabilized flames in vitiated flow. *Combust. Flame* **2012**, *159*, 291–305. [CrossRef]
4. Nemitallah, M.A.; Rashwan, S.S.; Mansir, I.B.; Abdelhafez, A.A.; Habib, M.A.M. Review of Novel Combustion Techniques for Clean Power Production in Gas Turbines. *Energy Fuels* **2018**, *32*, 979–1004. [CrossRef]
5. Fu, Q.; Kansha, Y.; Song, C.; Liu, Y.; Ishizuka, M.; Tsutsumi, A. A cryogenic air separation process based on self-heat recuperation for oxy-combustion plants. *Appl. Energy* **2016**, *162*, 1114–1121. [CrossRef]
6. Rashwan, S.S.; Abouarab, T.W. Experimental Investigation of Oxy-fuel combustion of CNG Flames Stabilized over A Perforated Plate Burner. In Proceedings of the 18th IFRF Members' Conference—Flexible and Clean Fuel Conversion to Industry, Freising, Germany, 1–3 June 2015; p. 25.
7. Jonsson, M.; Yan, J. Humidified gas turbines—A review of proposed and implemented cycles. *Energy* **2005**, *30*, 1013–1078. [CrossRef]
8. Giampaolo, T. *The Gas Turbine Handbook: Principles and Practices*; The Fairmont Press, Inc.: Gistrup, Denmark, 2003.
9. He, Y.; Zou, C.; Song, Y.; Chen, W.; Jia, H.; Zheng, C. Experimental and Numerical Study of the Effect of High Steam Concentration on the Oxidation of Methane and Ammonia during Oxy-Steam Combustion. *Energy Fuels* **2016**, *30*, 6799–6807. [CrossRef]
10. Richards, G.A.; Casleton, K.H.; Chorpening, B.T. CO_2 and H_2O diluted oxy-fuel combustion for zero-emission power. *Proc. Inst. Mech. Eng. Part A J. Power Energy* **2005**, *219*, 121–126. [CrossRef]
11. Jin, B.; Zhao, H.; Zou, C.; Zheng, C. Comprehensive investigation of process characteristics for oxy-steam com-bustion power plants. *Energy Convers. Manag.* **2015**, *99*, 92–101. [CrossRef]
12. Xiong, J.; Zhao, H.; Chen, M.; Zheng, C. Simulation Study of an 800 MWe Oxy-combustion Pulverized-Coal-Fired Power Plant. *Energy Fuels* **2011**, *25*, 2405–2415. [CrossRef]
13. Xiong, J.; Zhao, H.; Zheng, C. Exergy Analysis of a 600 MWe Oxy-combustion Pulverized-Coal-Fired Power Plant. *Energy Fuels* **2011**, *25*, 3854–3864. [CrossRef]
14. Xiong, J.; Zhao, H.; Zheng, C. Thermoeconomic cost analysis of a 600MWe oxy-combustion pulverized-coal-fired power plant. *Int. J. Greenh. Gas Control* **2012**, *9*, 469–483. [CrossRef]
15. Jin, B.; Zhao, H.; Zheng, C. Dynamic modeling and control for pulverized-coal-fired oxy-combustion boiler island. *Int. J. Greenh. Gas Control.* **2014**, *30*, 97–117. [CrossRef]
16. Singh, J. *Sterling Dictionary of Physics*; Sterling Publishers Pvt. Ltd.: New Delhi, India, 1999.

17. Shahsavan, M.; Mack, J.H. The effect of heavy working fluids on hydrogen combustion. In Proceedings of the 10th U.S. National Combustion Meeting, College Park, MD, USA, 23–26 April 2017.
18. No, H.-C.; Kim, J.-H.; Kim, H.-M. A review of helium gas turbine technology for high-temperature gas-cooled reactors. *Nucl. Eng. Technol.* **2007**, *39*, 21–30. [CrossRef]
19. Nayagam, V.; Haggard, J.B., Jr.; Colantonio, R.O.; Marchese, A.J.; Dryer, F.L.; Zhang, B.L.; Williams, F.A. Mi-crogravity n-heptane droplet combustion in oxygen-helium mixtures at atmospheric pressure. *AIAA J.* **1998**, *36*, 1369–1378. [CrossRef]
20. Li, H.; Yan, J.; Anheden, M. Impurity impacts on the purification process in oxy-fuel combustion based CO_2 capture and storage system. *Appl. Energy* **2009**, *86*, 202–213. [CrossRef]
21. Puebla, H.; Enrique, L.; Hurtado, J.I.L.; Soria, B.Y.M. Power cycle assessment of nuclear high temperature gas-cooled reactors. *Appl. Therm. Eng.* **2009**, *29*, 1759–1765.
22. Rao, A.B.; Rubin, E.S. A Technical, Economic, and Environmental Assessment of Amine-Based CO_2 Capture Technology for Power Plant Greenhouse Gas Control. *Environ. Sci. Technol.* **2002**, *36*, 4467–4475. [CrossRef]
23. Prisyazhniuk, V.A. Alternative trends in development of thermal power plants. *Appl. Therm. Eng.* **2008**, *28*, 190–194. [CrossRef]
24. Glarborg, P.; Bentzen, L.L.B. Chemical Effects of a High CO_2 Concentration in Oxy-Fuel Combustion of Methane. *Energy Fuels* **2007**, *22*, 291–296. [CrossRef]
25. Bibrzycki, J.; Poinsot, T. *Reduced Chemical Kinetic Mechanisms for Methane Combustion in O_2/N_2 and O_2/CO_2 Atmosphere;* Working note ECCOMET WN/CFD/10 17; CERFACT: Toulouse, France, 2010.
26. Al-Tamreh, S.A.; Ibrahim, M.H.; El-Naas, M.H.; Vaes, J.; Pant, D.; Benamor, A.; Amhamed, A. Electroreduction of Carbon Dioxide into Formate: A Comprehensive Review. *ChemElectroChem* **2021**, *8*, 3207–3220. [CrossRef]
27. Sa, D.A.; Al Zubaidy, S. Gas turbine performance at varying ambient temperature. *Appl. Therm. Eng.* **2011**, *31*, 2735–2739.
28. Pilavachi, P. Mini- and micro-gas turbines for combined heat and power. *Appl. Therm. Eng.* **2002**, *22*, 2003–2014. [CrossRef]
29. Heppenstall, T. Advanced gas turbine cycles for power generation: A critical review. *Appl. Therm. Eng.* **1998**, *18*, 837–846. [CrossRef]
30. ASME COGEN-TURBO. A Study on Modified Gas Turbine Systems with Steam Injection or Evaporative Regeneration. Annerwall, K., Svedberg, G., Eds.; ASME Turbo Expo Conference, FLORIDA, USA. 1991. Available online: https://www.researchgate.net/publication/282387540_A_study_on_modified_gas_turbine_systems_with_steam_injection_or_evaporative_regeneration (accessed on 6 January 2022).
31. Bhargava, R.; Meher-Homji, C.B. Parametric analysis of existing gas turbines with inlet evaporative and overspray fogging. In *ASME Turbo Expo 2002: Power for Land, Sea, and Air;* American Society of Mechanical Engineers: New York, NY, USA, 2002; Volume 36096, pp. 387–401.
32. McDonald, C.F.; Wilson, D.G. The utilization of recuperated and regenerated engine cycles for high-efficiency gas turbines in the 21st century. *Appl. Therm. Eng.* **1996**, *16*, 635–653. [CrossRef]
33. Al-Doboon, A.; Gutesa, M.; Valera-Medina, A.; Syred, N.; Ng, J.-H.; Chong, C.T. CO_2-argon-steam oxy-fuel (CARSOXY) combustion for CCS inert gas atmospheres in gas turbines. *Appl. Therm. Eng.* **2017**, *122*, 350–358. [CrossRef]
34. Luo, X.; Wang, M. Optimal operation of MEA-based post-combustion carbon capture for natural gas combined cycle power plants under different market conditions. *Int. J. Greenh. Gas Control.* **2016**, *48*, 312–320. [CrossRef]
35. Kanniche, M.; Gros-Bonnivard, R.; Jaud, P.; Valle-Marcos, J.; Amann, J.M.; Bouallou, C. Pre-combustion, post-combustion and oxy-combustion in thermal power plant for CO_2 capture. *Appl. Therm. Eng.* **2010**, *30*, 53–62. [CrossRef]
36. Altamash, T.; Amhamed, A.; Aparicio, S.; Atilhan, M. Effect of hydrogen bond donors and acceptors on CO_2 absorption by deep eutectic solvents. *Processes* **2020**, *8*, 1533. [CrossRef]
37. Pugh, D.; Bowen, P.; Valera-Medina, A.; Giles, A.; Runyon, J.; Marsh, R. Influence of steam addition and elevated ambient conditions on NOx reduction in a staged premixed swirling NH_3/H_2 flame. *Proc. Combust. Inst.* **2019**, *37*, 5401–5409. [CrossRef]
38. Hussein, N.A.; Valera-Medina, A.; Alsaegh, A.S. Ammonia- hydrogen combustion in a swirl burner with reduction of NOx emissions. *Energy Procedia* **2019**, *158*, 2305–2310. [CrossRef]
39. Smith, A.R.; Klosek, J. A review of air separation technologies and their integration with energy conversion processes. *Fuel Process. Technol.* **2001**, *70*, 115–134. [CrossRef]
40. Najjar, Y.S. Gas turbine cogeneration systems: A review of some novel cycles. *Appl. Therm. Eng.* **2000**, *20*, 179–197. [CrossRef]
41. Terry, F.W. Combustion processes for carbon capture. *Proc. Combust. Inst.* **2007**, *31*, 31–47.
42. Liang, Z.; Rongwong, W.; Liu, H.; Fu, K.; Gao, H.; Cao, F.; Zhang, R.; Sema, T.; Henni, A.; Sumon, K.Z.; et al. Recent progress and new developments in post-combustion carbon-capture technology with amine based solvents. *Int. J. Greenh. Gas Control* **2015**, *40*, 26–54. [CrossRef]
43. Kaur, M. Carbon Capturing and Storage Technology & current CCS initiatives in India (Emerging Technology in the field of Environmental Engineering). *Int. J. Adv. Res. Comput. Sci.* **2017**, *8*, 52–55.
44. Leung, D.Y.C.; Caramanna, G.; Maroto-Valer, M.M. An overview of current status of carbon dioxide capture and storage technologies. *Renew. Sustain. Energy Rev.* **2014**, *39*, 426–443. [CrossRef]
45. Stanger, R.; Wall, T. Sulphur impacts during pulverised coal combustion in oxy-fuel technology for carbon capture and storage. *Prog. Energy Combust. Sci.* **2011**, *37*, 69–88. [CrossRef]
46. Aaron, D.; Tsouris, C. Separation of CO_2 from Flue Gas: A Review. *Sep. Sci. Technol.* **2005**, *40*, 321–348. [CrossRef]

47. Knapik, E.; Chruszcz-Lipska, K. Chemistry of Reservoir Fluids in the Aspect of CO_2 Injection for Selected Oil Reservoirs in Poland. *Energies* **2020**, *13*, 6456. [CrossRef]
48. Rasmussen, L.; Fan, T.; Rinehart, A.; Luhmann, A.; Ampomah, W.; Dewers, T.; Heath, J.; Cather, M.; Grigg, R. Carbon Storage and Enhanced Oil Recovery in Pennsylvanian Morrow Formation Clastic Reservoirs: Controls on Oil–Brine and Oil–CO_2 Relative Permeability from Diagenetic Heterogeneity and Evolving Wettability. *Energies* **2019**, *12*, 3663. [CrossRef]
49. Cao, C.; Liu, H.; Hou, Z.; Mehmood, F.; Liao, J.; Feng, W. A Review of CO_2 Storage in View of Safety and Cost-Effectiveness. *Energies* **2020**, *13*, 600. [CrossRef]
50. Fawwaz Alrebei, O.; Al-Doboon, A.; Bowen, P.; Valera Medina, A. CO_2-Argon-Steam Oxy-Fuel Production for (CARSOXY) Gas Turbines. *Energies* **2019**, *12*, 3580. [CrossRef]
51. Amhamed, A.; Atilhan, M.; Berdiyorov, G. Permeabilities of CO_2, H_2S and CH_4 through choline-based ionic liquids: Atomistic-scale simulations. *Molecules* **2019**, *24*, 2014. [CrossRef] [PubMed]
52. Essehli, R.; Sabri, S.; El-Mellouhi, F.; Aïssa, B.; Yahia, H.B.; Altamash, T.; Khraisheh, M.; Amhamed, A.; El Bali, B. Single crystal structure, vibrational spectroscopy, gas sorption and antimicro-bial properties of a new inorganic acidic diphosphates material $(NH_4)_2Mg(H_2P_2O_7)_2 \bullet 2H_2O$. *Sci. Rep.* **2020**, *10*, 1–14. [CrossRef]
53. Cheung, H. Moderate-pressure cryogenic air separation process. *Gas Sep. Purif.* **1991**, *5*, 25–28. [CrossRef]
54. Khalel, Z.A.M.; Rabah, A.A.; Barakat, T.A.M. A New Cryogenic Air Separation Process with Flash Separator. *ISRN Thermodyn.* **2013**, *2013*, 1–4. [CrossRef]
55. Leo, A.; Liu, S.; da Costa, J.C.D. Development of mixed conducting membranes for clean coal energy delivery. *Int. J. Greenh. Gas Control* **2009**, *3*, 357–367. [CrossRef]
56. Kneer, R.; Toporov, D.; Förster, M.; Christ, D.; Broeckmann, C.; Pfaff, E.; Zwick, M.; Engels, S.; Modigell, M. OXYCOAL-AC: Towards an integrated coal-fired power plant process with ion transport membrane-based oxygen supply. *Energy Environ. Sci.* **2010**, *3*, 198–207. [CrossRef]
57. Sundkvist, S.G.; Julsrud, S.; Vigeland, B.; Naas, T.; Budd, M.; Leistner, H.; Winkler, D. Development and testing of AZEP reactor components. *Int. J. Greenh. Gas Control.* **2007**, *1*, 180–187. [CrossRef]
58. Wilson, J.; Christie, M.; Degenstein, N.; Shah, M.; Li, J. OTM based oxy-fuel combustion for CO_2 capture. In Proceedings of the 34th International Technical Conference on Clean Coal & Fuel Systems, Clearwater, FL, USA, 31 May–4 June 2009; Volume 31.
59. Holmes, M.J.; Ohrn, T.R.; Chen, C.M. Ion Transport Membrane Module and Vessel System with Directed Internal Gas Flow. U.S. Patent 7,658,788, 9 February 2010.
60. Den Exter, M.J.; Haije, W.G.; Vente, J.F. Viability of ITM technology for oxygen production and oxidation processes: Material, system, and process aspects. In *Inorganic Membranes for Energy and Environmental Applications*; Springer: New York, NY, USA, 2009; pp. 27–51.
61. Beggel, F.; Engels, S.; Modigell, M.; Nauels, N. *Oxyfuel Combustion by Means of High Temperature Membranes for Air Separation*; U.S. Department of Energy: Hyogo, Japan, 2009.
62. Stadler, H.; Beggel, F.; Habermehl, M.; Persigehl, B.; Kneer, R.; Modigell, M.; Jeschke, P. Oxyfuel coal combustion by efficient integration of oxygen transport membranes. *Int. J. Greenh. Gas Control* **2011**, *5*, 7–15. [CrossRef]
63. Christou, C.; Hadjipaschalis, I.; Poullikkas, A. Assessment of integrated gasification combined cycle technology competitiveness. *Renew. Sustain. Energy Rev.* **2008**, *12*, 2459–2471. [CrossRef]
64. Ordorica-Garcia, G.; Douglas, P.; Croiset, E.; Zheng, L. *Techno-Economic Evaluation of IGCC Power Plants with CO_2 Capture*; U.S. Department of Energy: Hyogo, Japan, 2005.
65. Bohm, M.C.; Herzog, H.J.; Parsons, J.E.; Sekar, R.C. Capture-ready coal plants—Options, technologies and economics. *Int. J. Greenh. Gas Control.* **2007**, *1*, 113–120. [CrossRef]
66. Guocai, T. Applications of green solvents in toxic gases removal. In *Green Sustainable Process for Chemical and Environmental Engineering and Science*; Elsevier: Amsterdam, The Netherlands, 2021; pp. 149–201.
67. Song, Z.Y.; Wang, B.-W.; Song, K.; Zheng, C.-G. The performance research on new oxygen carrier CASO_4 used in chemical-looping combustion. *J. Eng. Thermophys.* **2006**, *13*, 3.
68. Abad, A.; Mattisson, T.; Lyngfelt, A.; Rydèn, M. Chemical-looping combustion in a 300 W continuously operating reactor system using a manganese-based oxygen carrier. *Fuel* **2006**, *85*, 1174–1185. [CrossRef]
69. Mattisson, T.; Johansson, M.; Lyngfelt, A. The use of NiO as an oxygen carrier in chemical-looping combustion. *Fuel* **2006**, *85*, 736–747. [CrossRef]
70. Abad, A.; Adánez, J.; García-Labiano, F.; Luis, F.; Gayán, P.; Celaya, J. Mapping of the range of operational conditions for Cu-, Fe-, and Ni-based oxygen carriers in chemical-looping combustion. *Chem. Eng. Sci.* **2007**, *62*, 533–549. [CrossRef]
71. Abad, A.; Mattisson, T.; Lyngfelt, A.; Johansson, M. The use of iron oxide as oxygen carrier in a chemical-looping reactor. *Fuel* **2007**, *86*, 1021–1035. [CrossRef]
72. Agrawal, R.; Woodward, D.W.; Yee, T.F. Argon production from air distillation: Use of a heat pump in a ternary distillation with a side rectifier. *Gas Sep. Purif.* **1994**, *8*, 37–43. [CrossRef]
73. Cornelissen, R.; Hirs, G. Exergy analysis of cryogenic air separation. *Energy Convers. Manag.* **1998**, *39*, 1821–1826. [CrossRef]
74. Barelli, L.; Bidini, G.; Gallorini, F.; Servili, S. Hydrogen production through sorption-enhanced steam methane reforming and membrane technology: A review. *Energy* **2008**, *33*, 554–570. [CrossRef]

75. LeValley, T.L.; Richard, A.R.; Fan, M. The progress in water gas shift and steam reforming hydrogen production technologies—A review. *Int. J. Hydrog. Energy* **2014**, *39*, 16983–17000. [CrossRef]
76. Lee, F.B. A comparative study of fuels for on-board hydrogen production for fuel-cell-powered automo-biles. *Int. J. Hydrog. Energy* **2001**, *26*, 381–397.
77. Sjardin, M.; Damen, K.J.; Faaij, A.P.C. Techno-economic prospects of small-scale membrane reactors in a future hydrogen-fuelled transportation sector. *Energy* **2006**, *31*, 2523–2555. [CrossRef]
78. Khzouz, M.; Gkanas, E.I. Experimental and Numerical Study of Low Temperature Methane Steam Reforming for Hydrogen Production. *Catalysts* **2017**, *8*, 5. [CrossRef]
79. Ballard. Available online: http://www.ballard.com/ (accessed on 30 April 2020).
80. Martyn, V.T. *Catalyst Handbook*; Routledge: New York, NY, USA, 2018.
81. Halabi, M.H.; de Croon, M.H.J.M.; van der Schaaf, J.; Cobden, P.D.; Schouten, J.C. Intrinsic kinetics of low tem-perature catalytic methane–steam reforming and water–gas shift over Rh/CeαZr1− αO2 catalyst. *Appl. Catal. A Gen.* **2010**, *389*, 80–91. [CrossRef]
82. Angeli, S.; Monteleone, G.; Giaconia, A.; Lemonidou, A. State-of-the-art catalysts for CH4 steam reforming at low temperature. *Int. J. Hydrog. Energy* **2014**, *39*, 1979–1997. [CrossRef]
83. Nieva, M.A.; Villaverde, M.M.; Monzón, A.; Garetto, T.F.; Marchi, A.J. Steam-methane reforming at low temperature on nickel-based catalysts. *Chem. Eng. J.* **2014**, *235*, 158–166. [CrossRef]
84. Santo, V.D.; Gallo, A.; Naldoni, A.; Guidotti, M.; Psaro, R. Bimetallic heterogeneous catalysts for hydrogen production. *Catal. Today* **2012**, *197*, 190–205. [CrossRef]
85. Andersson, M.; Paradis, H.; Yuan, J.; Sundén, B. Review of catalyst materials and catalytic steam reforming reactions in SOFC anodes. *Int. J. Energy Res.* **2011**, *35*, 1340–1350. [CrossRef]
86. Li, D.; Nakagawa, Y.; Tomishige, K. Methane reforming to synthesis gas over Ni catalysts modified with noble metals. *Appl. Catal. A Gen.* **2011**, *408*, 1–24. [CrossRef]
87. Sircar, S.; Golden, T.C. Pressure swing adsorption technology for hydrogen production. *Hydrog. Syngas Prod. Purif. Technol.* **2009**, *10*, 414–450.
88. Smith, R.J.B.; Loganathan, M.; Shantha, M.S. A review of the water gas shift reaction kinetics. *Int. J. Chem. React. Eng.* **2010**, *8*, 11–44. [CrossRef]
89. George, W.; Dresselhaus, M.S.; Buchanan, M.V. The hydrogen economy. *Phys. Today* **2004**, *57*, 39–44.
90. Osvaldo, V.-Z.M.; Nicholas, S.; Agustín, V.-M.; Daniel, D.L.R.-U. Flashback Avoidance in Swirling Flow Burners. *Ing. Investig. Tecnol.* **2014**, *15*, 603–614. [CrossRef]
91. Tian, Z.; Jiang, B.; Malik, A.; Zheng, Q. Axial helium compressor for high-temperature gas-cooled reactor: A review. *Ann. Nucl. Energy* **2019**, *130*, 54–68. [CrossRef]
92. Kim, J.H.; No, H.C.; Kim, H.M.; Lim, H.S. Direct implementation of an axial-flow helium gas turbine tool in a system analysis tool for HTGRs. *Nucl. Eng. Des.* **2008**, *238*, 3379–3388. [CrossRef]
93. Muto, Y.; Ishiyama, S.; Fukuyama, Y.; Okumoto, J.; Kishibe, T.; Yamada, S. Design Study of Helium Turbine for the 300 MW HTGR-GT Power Plant. In *Turbo Expo: Power for Land, Sea, and Air*; American Society of Mechanical Engineers: New York, NY, USA, 2000; Volume 78552, p. V002T04A010.
94. Tian, Z.; Zheng, Q.; Malik, A.; Jiang, B. Numerical investigation of the effect of highly loaded design on the tip leakage in helium compressor rotors. *Prog. Nucl. Energy* **2018**, *105*, 263–270. [CrossRef]
95. Elistratov, S.L.; Vitovskii, O.V.; Slesareva, E.Y. Experimental investigation of heat transfer of helium-xenon mixtures in cylindrical channels. *J. Eng. Thermophys.* **2015**, *24*, 33–35. [CrossRef]
96. Vitovsky, O.; Nakoryakov, V.E.V.; Slesareva, E.Y. Heat transfer of helium–xenon mixture on the initial pipe section. *J. Eng. Thermophys.* **2015**, *24*, 338–341. [CrossRef]
97. Nakoryakov, V.E.; Vitovsky, O.V. Study of heat transfer of a helium–xenon mixture in heated channels with different cross-sectional shapes. *J. Appl. Mech. Tech. Phys.* **2017**, *58*, 664–669. [CrossRef]
98. Dragunov, Y.G.; Smetannikov, V.P.; Gabaraev, B.A.; Orlov, A.N.; Belyakov, M.S.; Derbenev, D.S. On calculation of the transport coefficients and thermodynamic properties of a helium-xenon gas mixture. *J. Eng. Thermophys.* **2013**, *22*, 21–29. [CrossRef]
99. Haire, M.A.; Vargo, D.D. Review of Helium and Xenon Pure Component and Mixture Transport Properties and Recommendation of Estimating Approach for Project Prometheus (Viscosity and Thermal Conductivity). In *AIP Conference Proceedings*; American Institute of Physics: College Park, MD, USA, 2007; Volume 880, pp. 559–570. [CrossRef]

separations

Article

Quantitation of Acetyl Hexapeptide-8 in Cosmetics by Hydrophilic Interaction Liquid Chromatography Coupled to Photo Diode Array Detection

Vasiliki Raikou [1], Eleni Kalogria [1], Athanasia Varvaresou [2], Efthimios Tsirivas [2] and Irene Panderi [1,*]

[1] Laboratory of Pharmaceutical Analysis, Department of Pharmacy, Division of Pharmaceutical Chemistry, National and Kapodistrian University of Athens, Panepistimipolis-Zografou, 15771 Athens, Greece; vasraikou@gmail.com (V.R.); ekalogria@gmail.com (E.K.)

[2] Laboratory of Chemistry-Biochemistry-Cosmetic Science, Department of Biomedical Sciences, Faculty of Health and Welfare Sciences, Campus 1, University of West Attica, Agiou Spyridonos, Egaleo, 12243 Athens, Greece; avarvares@uniwa.gr (A.V.); tsirivas@cellco.gr (E.T.)

* Correspondence: irenepanderi@gmail.com; Tel.: +30-210-727-4820

Abstract: Bioactive peptides are gaining more and more popularity in the research and development of cosmetic products with anti-aging effect. Acetyl hexapeptide-8 is a hydrophilic peptide incorporated in cosmetics to reduce the under-eye wrinkles and the forehead furrows. Hydrophilic interaction liquid chromatography (HILIC) is the separation technique of choice for analyzing peptides. In this work, a rapid HILIC method coupled to photodiode array detection operated at 214 nm was developed, validated and used to determine acetyl-hexapeptide-8 in cosmetics. Chromatography was performed on a Xbridge® HILIC BEH analytical column using as mobile phase a 40 mM ammonium formate water solution (pH 6.5)-acetonitrile mixture 30:70, *v/v* at flow rate 0.25 mL min^{-1}. The assay was linear over the concentration range 20 to 30 μg mL^{-1} for the cosmetic formulations and 0.004 to 0.007% (*w/w*) for the cosmetic cream. The limits of quantitation for acetyl hexapeptide-8 were 1.5 μg mL^{-1} and 0.002% (*w/w*) for the assay of cosmetic formulations and cosmetic creams, respectively. The method was applied to the analysis of cosmetic formulations and anti-wrinkle cosmetic creams.

Keywords: hydrophilic interaction liquid chromatography; chromatography; oligopeptides; acetyl hexapeptide-8; acetyl hexapeptide-3; Argireline; cosmetics

Citation: Raikou, V.; Kalogria, E.; Varvaresou, A.; Tsirivas, E.; Panderi, I. Quantitation of Acetyl Hexapeptide-8 in Cosmetics by Hydrophilic Interaction Liquid Chromatography Coupled to Photo Diode Array Detection. *Separations* **2021**, *8*, 125. https://doi.org/10.3390/separations8080125

Academic Editor: Alberto Cavazzini

Received: 31 July 2021
Accepted: 17 August 2021
Published: 18 August 2021

1. Introduction

Skin aging is a biological process *influenced* by various genetic, environmental (pollution, UV radiation), hormonal and metabolic factors. For more than 20 years, bioactive ingredients have been incorporated in cosmetics to smooth out deep wrinkles, improve skin elasticity and reduce the effects of the skin aging [1]. Given that most natural processes within the body are stimulated through the interaction of peptides and proteins with their target partners, bioactive peptides incorporated in cosmetics are one of the most popular ways to reduce wrinkles and fine lines, improve skin appearance and texture, and treat decolorated skin [2]. The role of peptides is crucial in several natural processes related to skin care, such as the modulation of cell proliferation, inflammation, melanogenesis, cell migration, angiogenesis and the synthesis and regulation of proteins. Bioactive peptides are gaining more and more popularity in the research and development of cosmetic products with anti-aging effect [3]. Acetyl hexapeptide-8, also known as acetyl-hexapeptide-3, is a neurotransmitter inhibitor peptide designed from the N-terminal end of the synaptosomal-associated protein (SNAP25) [4,5]. It competitive inhibits the SNAP25 component of the said vesicle docking and fusion protein complex (SNARE) [6] which triggers the calcium- dependent neurotransmitter release into the synapse, a process necessary for muscle contraction [7–10]. Acetyl hexapeptide-8 is marketed as Argireline® [11], and it has been efficiently used in cosmetics for smoothening the under-eye wrinkles and the forehead furrows [12–14]. After topical application at specific areas of the face, it inhibits the reactions that cause muscles to move or contract for example when forming facial expressions such as smiling or frowning [15,16]. A clinical trial of daily topical application of acetyl hexapeptide-8 in 24 patients with blepharospasm concluded that topical application of this peptide is safe and promising for prolonging the action of injectable botulinum neurotoxin therapy [17]. The quality control of cosmetic products containing bioactive peptides should be addressed under a more systematic investigation and the concentration of these peptides in cosmetics should be supported by product-specific studies. Therefore, there is a real need to set up analytical methods to quantitate the bioactive compounds in cosmetic products.

In the last decade, the increased interest in the separation of peptides has gained momentum due to the emphasis on the chromatographic separation of various proteins in proteome. High performance liquid chromatography (HPLC) has been widely used in the analysis of peptides in various fields of research and development, using different modes of separation. Nowadays, hydrophilic interaction liquid chromatography is the separation technique of choice for the analysis of peptides [18–22]. Usually, bioactive peptides incorporated in cosmetic products are hydrophilic compounds that show little or no retention on conventional RP-HPLC analytical columns. The stationary phases in HILIC are mainly polar such as silica gel, diol-, amino- or cyano- bonded and other zwitterionic packing materials and the typical mobile phases consists of mixtures of a highly polar solvent (typically water) with an organic modifier (typically acetonitrile) [23,24]. The analytes retention is increased by increasing the proportion of the organic modifier in the mobile phase [25,26]. The polar functional groups on the HILIC stationary phases absorb water (0.5%–1.0%) and this way a water-enriched layer is immobilized between the mobile and the stationary phase, especially when the water content of the mobile phase is less than 40% [27]. In HILIC the more hydrophilic analytes are eluted later than the less polar compound. In the literature several works have been published on peptide separation using HILIC columns [28–30], but only a few studies were dedicated to peptide quantitation in cosmetics [31–33]. In one of these publications, ultra-high performance liquid chromatography coupled to tandem mass spectrometry (UPLC-MS/MS) using a TSK-gel Amide-80® HILIC analytical column was used to quantitate acetyl hexapeptide-8 in cosmetics after solid phase extraction procedure [34]. The TSK-gel Amide-80® HILIC analytical column was also used in LC-MS/MS methods developed to evaluate transdermal absorption of acetyl-hexapeptide-8 [35,36]. Even though the abovementioned LC-MS/MS approaches are selective and sensitive for the quantitation of acetyl-hexapeptide-8 in various matrices, there is a real need for the development of a reliable analytical method without the need for specialized equipment that can be used in routine analyses.

In this work a rapid and sensitive hydrophilic interaction liquid chromatographic method with photodiode array (PDA) detection was developed and validated for the quantitation of acetyl-hexapeptide-8 in cosmetics. The appropriate stationary phase, pH, buffer concentration and mobile phase composition, was thoroughly investigated prior to method validation. The combination of HILIC with PDA detection provides an accurate, repeatable and robust method for the quantitative analysis of cosmetic products. To the best of our knowledge, this is the first report of a HILIC-PDA method for the quantitation of acetyl-hexapeptide-8 in cosmetics. In this work a Xbridge®-HILIC BEH analytical column has been used and combined with a rapid and simple sample pretreatment. All the above in combination with a short run time of less than 10 min, makes the proposed HILIC-PDA method suitable for the routine quality control of cosmetics.

2. Materials and Methods

2.1. Chemical and Reagents

HPLC grade solvents were bought from Sigma-Aldrich (St. Louis, MO, USA). Analytical reagent grade ammonium formate was acquired from Alfa-Aesar (Haverhill, MA,

USA). HPLC grade water was produced by means of a Synergy® UV water purification system (Merck Millipore, MA, USA). Whatman nylon membrane filters with pore size 0.45 μm and diameter 47 mm were purchased from GelmanSciences (Northampton, UK). Hydrophobic polytetrafluorethylene (PTFE phobic 13 mm, pore size 0.22 μm) syringe filters were acquired from Novalab SA, Athens, Greece representative of RephiLe Bioscience Ltd., Europe.

Acetyl hexapeptide-8 (97% pure), acetyl-Glu-Glu-Met-Gln-Arg-ArgNH2 was kindly provided from Cellco Chemicals Ltd. (Athens, Greece) distributor in Greece of Caregen Co. Ltd. (Gyeonggi, Korea).

Acetyl hexapeptide-8 formulation, namely Argireline peptide solution C®, was donated from N. Krallis S.A. distributor in Greece of Lipotec S.A., Spain containing aqua, 0.05% *w/w* acetyl hexapeptide-8 and 0.05% *w/w* 1,2 octanediol.

The anti-wrinkle cosmetic cream containing 0.005% *w/w* acetyl hexapeptide-8 was produced in the Laboratory of Chemistry-Biochemistry-Cosmetic Science, Department of Biomedical Sciences, University of West Attica in Athens, Greece by using the Argireline peptide solution C®. The excipients present in anti-wrinkle cream are aqua, xalifin-15, propylene glycol, sabowax FX-65, squalene, butylated hydroxyl toluene (BHT) and germall 115. Placebo cream containing only the excipients, without acetyl hexapeptide-8, was also prepared for validation purposes.

2.2. Stock and Calibration Standard Solutions

Acetyl hexapeptide-8 stock standard solution at 500 μg mL^{-1} was prepared in acetonitrile-water mixture (60:40, *v/v*). The stock standard solution was further diluted in acetonitrile-water mixture (60:40, *v/v*) to prepare mixed working standard solutions. The solutions were stored in amber bottles at 4 °C and remained stable for two months.

For the quantitation of acetyl hexapeptide-8 in Argireline peptide solution C® (formulation), the calibration samples were prepared at concentration levels 20, 23, 25, 28 and 30 μg mL^{-1} in water/acetonitrile mixture (30:70, *v/v*). Quality control (QC) samples were also prepared at concentration levels 20 and 25 μg mL^{-1}.

For the quantitation of acetyl hexapeptide-8 in the anti-wrinkle cosmetic cream the calibration spiked cream samples at concentration levels 0.004, 0.0045, 0.005, 0.006 and 0.007 *w/w* were prepared by spiking placebo cream with appropriate dilutions of acetyl hexapeptide-8 stock standard solution. Quality control (QC) samples were also prepared in a similar manner at concentration levels 0.004, 0.005 and 0.007% *w/w*.

2.3. Sample Preparation Procedure

2.3.1. Cosmetic Formulation

An accurately weighted amount (0.5 g) of Argireline peptide solution C® is transferred at a 10 mL volumetric flask and diluted to volume with water/acetonitrile (30:70, *v/v*). A portion of this solution is then analyzed by the proposed HILIC-PDA method for the quantitation of acetyl hexapeptide-8.

2.3.2. Cosmetic Cream

An accurately weighted amount (0.1 g) of cosmetic cream is mixed with 200 μL of acetonitrile-water mixture (60:40, *v/v*) and 800 μL of 30% 40 mM ammonium formate water solution in acetonitrile. The mixture is shaken for 2 min and then centrifuged at 18,000× *g* for 30 min, at ambient temperature. The upper layer is then filtered through a PTFE hydrophobic syringe filter prior to HILIC-PDA analysis.

2.4. HILIC-PDA

The HPLC-PDA analytical instrument used in this work is consisting of a Waters 717 plus autosampler, a column temperature oven, a Waters 1515 isocratic pump and a Waters 996 photodiode array detector (Milford, MA, USA). The Empower software (Milford, MA, USA) was used for the data acquisition. The chromatographic eluent was monitored over the wavelengths 200 to 400 nm and extracted chromatograms at λ 214 nm were used for data analysis. An Xbridge®-HILIC BEH guard column (20 × 2.1 mm, 3.5 μm) in line with an Xbridge®-HILIC BEH analytical column (2.1 × 150 mm, 3.5 μm) were used for the chromatography. The mobile phase was composed of 30% 40 mM ammonium formate aqueous solution in acetonitrile and pumped at a flow rate of 0.25 mL min^{-1}. Prior to the chromatography the mobile phase was filtered through a 0.22 μm nylon membrane filter, Membrane Solutions (Kent, WA, USA) and degassed under vacuum. Samples were injected via a 10 μL injection loop and acetyl hexapeptide-8 was quantitated in cosmetic products with a chromatographic run time of 10 min.

2.5. Method Validation and Application to the Analysis of Real Samples

The HILIC-PDA method was validated in terms of linearity, limit of detection, limit of quantitation, intra-day and inter-day precision and overall accuracy [37]. The method was applied to the analysis of various batches of a cosmetic formulation namely Argireline peptide solution C® and to the analysis of various batches of an anti-wrinkle cosmetic cream. To evaluate the linearity, linear regressions were used to construct the calibration graphs after the analysis calibration standards and spiked cream samples at five different concentration levels. Peak area measurements were used for quantitation of acetyl hexapeptide-8. The % coefficient of variations (%CVs) and the % relative recovery were calculated to evaluate precision (intra- and inter-day) and overall accuracy, respectively.

3. Results and Discussions

3.1. HILIC-PDA Method Development

Chromatography

Acetyl-hexapeptide-8 consists of a six amino acids chain acetylated at the N-terminal residue, N-acetyl-Glu-Glu-Met-Gln-Arg-ArgNH$_2$, as shown in Figure 1a. The ionization fraction and the LogD values of acetyl-hexapeptide-8 as a function of pH were calculated by the ADME boxes software ver. 3.0.3 built 45, Pharma Algorithms Ltd. (Toronto Canada) and the results are presented in Figure 1b.

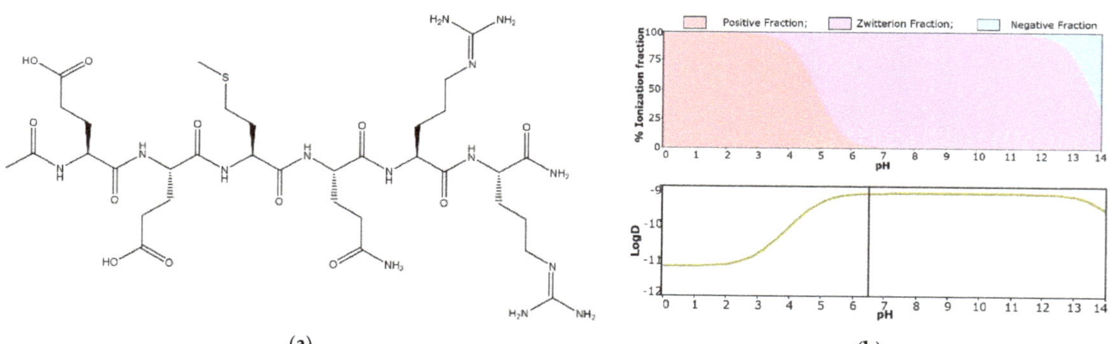

(a) (b)

Figure 1. Acetyl hexapeptide-8 (**a**) Chemical structure; (**b**) Plots of the ionization fraction (**top**) and the LogD values (**bottom**) as a function of pH.

LogD values of acetyl hexapeptide-8 versus pH are less than -9, indicating that this compound is highly hydrophilic (Figure 1b, bottom). HILIC is the analytical technique of choice for the chromatographic analysis of polar compounds, and it was therefore used in the present work. The retention mechanism in HILIC is based on several types of interactions such as adsorption, partition, electrostatic, hydrogen bonding and reversed-phase interactions [38–40]. Greater retention is achieved when more than 70% of organic modifier (e.g., acetonitrile) is used in the mobile phase. The chromatographic conditions were optimized to achieve adequate retention and optimum peak shape of acetyl-hexapeptide-8.

To find the optimal mobile phase composition we examined various combinations of aqueous buffer solutions and acetonitrile with changed content of each component. The detection wavelength was set to 214 nm, because at this wavelength acetyl heptpeptide-8 shows satisfactory absorption. The flow rate was set to 0.25 mL min^{-1} and the experiments were conducted at ambient temperature.

The XBridge®-HILIC BEH analytical column used in this work consists of BEH particles. Some accessible free silanol groups on the surface of these BEH particles are responsible for electrostatic interactions. The addition of aqueous salt solutions in the HILIC mobile phase eluents reduces the electrostatic interactions between the stationary phase and the analyte [41]. Plot of the logk values of acetyl hexapeptide-8 as a function of the concentration of ammonium formate is presented in Figure 2a. Ammonium formate concentration was modified from 2.5 to 100 mM with a constant aqueous component of the mobile phase eluent at 30%, *v/v* and a constant pH at 6.5. Under these conditions the free silanol groups on the surface of these BEH particles are negatively charged and acetyl hexapeptide-8 is in zwitterion form. The results showed that by increasing the concentration of ammonium formate the retention of acetyl hexapeptide-8 is slightly decreased up to 40 mM and then increased up to 100 mM. These findings indicate that the retention mechanism of acetyl hexapeptide-8 in XBridge®-HILIC BEH analytical column is complex and comprises both hydrophilic partition with secondary electrostatic interactions. From these experiments, we concluded that by using a 40 mM ammonium formate concentration in the mobile phase peak symmetry and plate numbers are improved and the analyte is adequately retained and well separated from the solvent front.

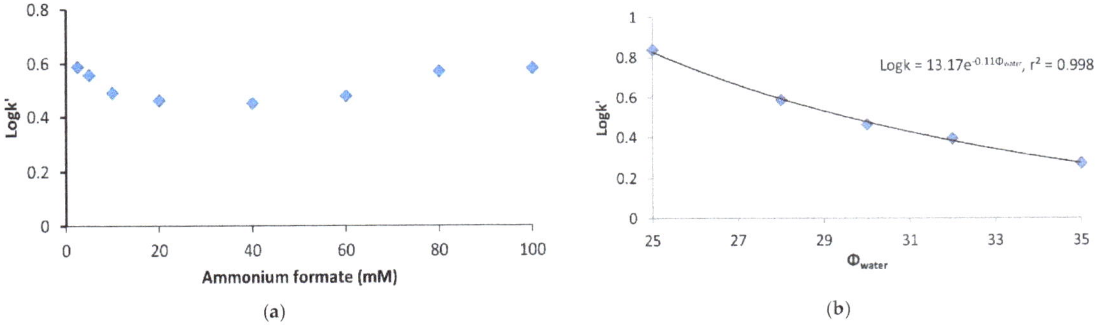

Figure 2. Plots of logk values for acetyl hexapeptide-8 as a function of: (**a**) the concentration of ammonium formate, and (**b**) the percentage of the water content of the mobile phase.

The chromatography of acetyl hexapeptide-8 was also explored by using various mobile phases where the concentration of ammonium formate in the whole mobile phase was kept constant at 12 mM, while the percentage of water, Φ_{water} varied from 25% to 35%. As shown in Figure 2b, the logk values of the peptide decrease exponentially with increasing Φ_{water}, implying a complex retention mechanism for this analyte on the selected HILIC column. The optimum mobile phase composition consists of 30% 40 mM ammonium formate water solution (pH 6.5) in acetonitrile. As shown in Figure 3, acetyl hexapeptide-8 is eluted at 8.15 min and the proposed HILIC-PDA method allows the isocratic elution of acetyl hexapeptide-8 within 10 min.

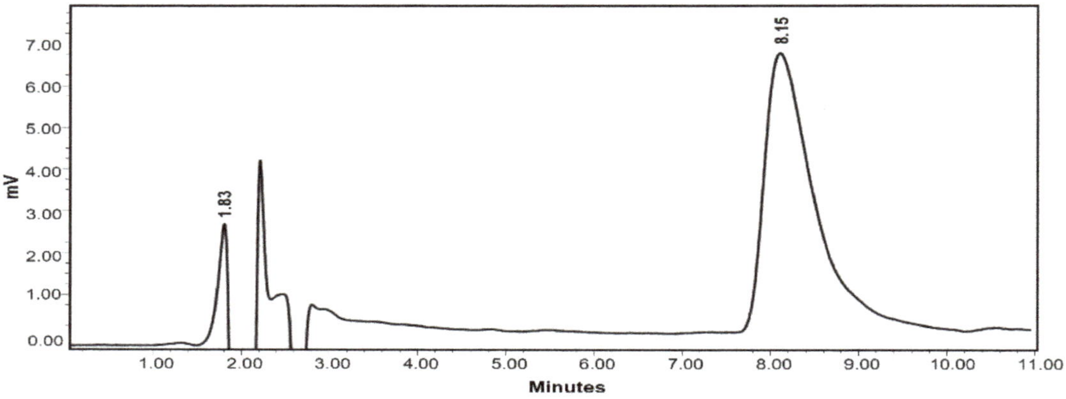

Figure 3. HPLC-PDA chromatogram of a quality control sample of acetyl hexapeptide-8 prepared in water/acetonitrile (30:70, *v/v*) at 25 µg mL^{-1}. Chromatographic conditions: BEH XBridge®-HILIC analytical column; mobile phase: 40 mM ammonium formate aqueous solution/acetonitrile (30:70, *v/v*), flow rate 0.25 mL min^{-1} at 214 nm.

3.2. Statistical Analysis of Data

3.2.1. Selectivity

The selectivity of the HILIC-PDA method to the analysis of Argireline peptide solution C® (cosmetic formulation) is demonstrated in Figure 4, where a chromatogram obtained from the analysis of the cosmetic formulation (red spiked line) is superimposed to a chromatogram of a quality control sample of acetyl hexapeptide-8 prepared in water/acetonitrile (30:70, *v/v*), both samples contain the analyte at 25 µg mL^{-1} (grey line).

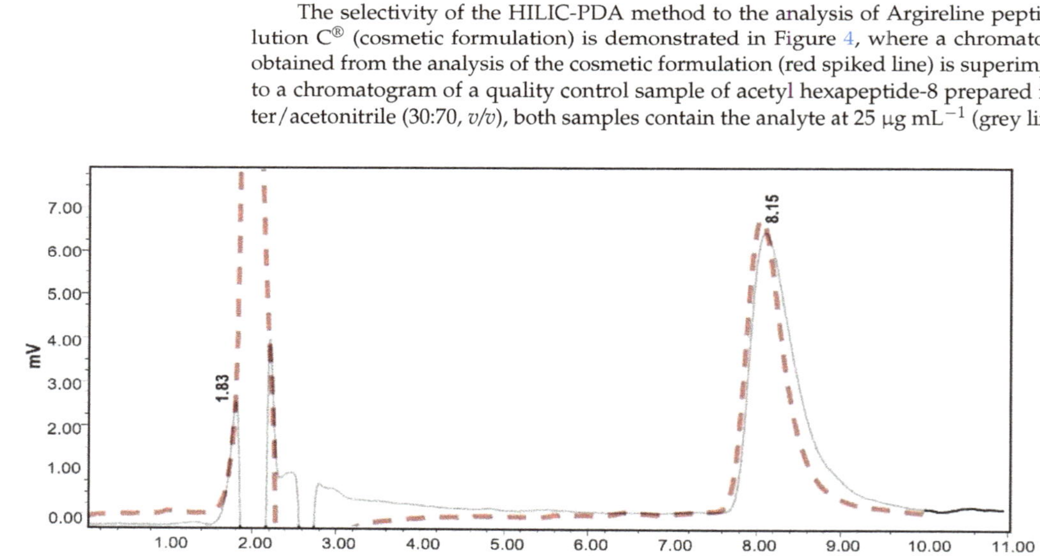

Figure 4. HPLC-PDA chromatogram obtained from the analysis of Argireline peptide solution C® (red spiked line) superimposed to a chromatogram of a quality control sample of acetyl hexapeptide-8 prepared in water/acetonitrile 30:70, *v/v* (grey line), both samples contain the analyte at 25 µg mL^{-1}. Chromatographic conditions: BEH XBridge®-HILIC analytical column; mobile phase: 40 mM ammonium formate aqueous solution/acetonitrile (30:70, *v/v*), flow rate 0.25 mL min^{-1} at 214 nm.

Moreover, the selectivity of the HILIC-PDA method to the analysis of cosmetic creams incorporated with acetyl hexapeptide-8 is demonstrated in Figure 5, where a chromatogram obtained from the analysis of a placebo cream sample (black line) is superimposed to a chromatogram of a cream sample obtained after the sample preparation described in Section 2.3.2 containing acetyl hexapeptide-8 at 0.005% *w/w* (blue line).

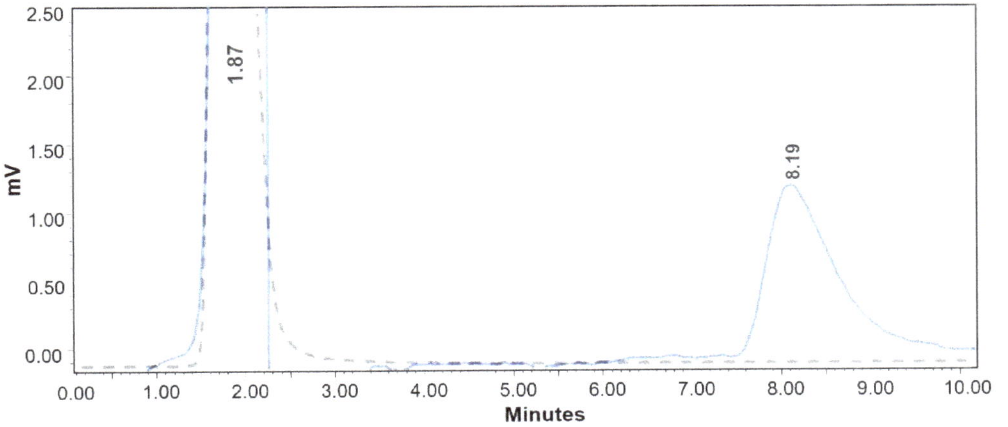

Figure 5. HPLC-PDA chromatogram obtained from the cosmetic cream (blue line) superimposed to a chromatogram of a black cream sample (grey spiked line). Chromatographic conditions: BEH XBridge®-HILIC analytical column; mobile phase: 40 mM ammonium formate aqueous solution/acetonitrile (30:70, *v/v*), flow rate 0.25 mL min^{-1} at 214 nm.

3.2.2. Linearity Data

For the quantitation of acetyl hexapeptide-8 in the cosmetic formulation (Argireline peptide solution C®) the calibration curves have been constructed at the range of concentrations 20 to 30 μg mL^{-1}. The peak area signal of the peptide, S, versus the corresponding concentrations, C exhibited linear relationships and the results of a typical calibration curve are shown in Table 1. A Student's *t*-test was also performed to evaluate whether the intercept of the regression equation was significantly different from the theoretical zero value. The test was based on the estimation of the experimental t-value, $t_{experimental} = a/Sa$, where a is the intercept and Sa is the standard deviation of the intercept of the regression equation, and on the comparison of $t_{experimental}$ with the theoretical t-value. tp. The results presented in Table 1 indicate that the intercept of the regression equation does not differ from the theoretical zero value.

Table 1. Linearity data for the quantitation of acetyl hexapeptide-8 in Argireline peptide solution C® (formulation) as assessed by the HILIC-PDA method.

Concentration Range (μg mL^{-1})	20–30
Regression Equation	$S = 18.08\ (\pm 0.64) \times 10^3\ C - 20\ (\pm 16) \times 10^3$
Correlation Coefficient, R	0.998
Standard Error of Estimation, Sr	5072
Limit of Detection, LOD (μg mL^{-1})	0.5
Limit of Quantitation, LOQ (μg mL^{-1})	1.5
$t_{experimental}$: a/S_a [1]	1.24
tp, f = 3, $p = 0.05$ [2]	4.30

[1] $t_{experimental}$ = experimental t-value; a = intercept; Sa standard deviation of the intercept; [2] tp = theoretical t-value; f = degrees of freedom; *p* = *p*-value.

For the quantitation of acetyl hexapeptide-8 in cosmetic creams, calibration curves were constructed after the analysis of spiked cream samples over the concentration range 0.004 to 0.007 *w/w*. The results of a typical calibration curve are presented in Table 2. In all cases correlation coefficient is greater than 0.994 indicating linear relationships between the peak area signal of the analyte, S and the corresponding concentrations, C. A Student's *t*-test was also performed in analogous manner, and the results (Table 2) indicate that the intercept of the regression line is not significantly different from zero and thus there is no interference from the cream matrix.

Table 2. Linearity data for the quantitation of acetyl hexapeptide-8 in cosmetic cream as assessed by the HILIC-PDA method.

Concentration Range (% *w/w*)	0.004–0.007
Regression Equation	S = 1105 (±41) × C + 0.71 (±0.23)
Correlation Coefficient, r	0.998
Standard Error of Estimation, Sr	0.098
Limit of Detection, LOD (% *w/w*)	6.8×10^{-4}
Limit of Quantitation, LOQ (% *w/w*)	0.002
$t_{experimental}$: a/S_a [1]	3.10
tp, f = 3, $p = 0.05$ [2]	4.30

[1] $t_{experimental}$ = experimental t-value; a = intercept; Sa standard deviation of the intercept; [2] tp = theoretical t-value; f = degrees of freedom; *p* = *p*-value.

Limit of detection (LOD) and limit of quantitation (LOQ) values for acetyl hexapeptide-8 were calculated as the amounts for which the signal-to-noise ratios were 3:1 and 10:1, respectively. This was achieved by the analysis of dilute solutions of the peptide at known concentration prepared by the appropriate sample preparation procedure [42]. LOD and LOQ values for acetyl hexapeptide-8 in cosmetic formulation and in cosmetic cream are reported in Table 1 and in Table 2, respectively.

3.2.3. Accuracy and Precision

Precision and accuracy were evaluated by one-way analysis of variance (ANOVA) and the results are presented in Table 3. The total precision was between 1.74 to 4.34 for the cosmetic formulation and 2.46 to 3.53% for acetyl hexapeptide-8 in cosmetic cream. The total accuracy was between 98.9 to 99.8% for the analyte in cosmetic formulation and 99.3 to 101.6% for the quantitation in cosmetic cream.

Table 3. Accuracy and precision data of the HILIC-PDA method for the quantitation of acetyl hexapeptide-8 in cosmetic formulation and cosmetic creams (*n* = 3 runs in 5 replicates).

Matrix	Concentration Levels		
Cosmetic Formulation			
Added concentration (μg mL^{-1})	20		25
Overall mean	19.95 (±0.91)		24.74 (±0.42)
Intraday CV(%)[1]	4.76		1.66
Total precision CV(%) [1]	4.37		1.74
Total accuracy Er%[2]	99.8		98.9
Cosmetic cream			
Added concentration (% *w/w*)	0.004	0.005	0.007
Overall mean	$39.7 (\pm 0.82) \times 10^{-4}$	$50.62 (\pm 0.39) \times 10^{-4}$	$69.71 (\pm 0.79) \times 10^{-4}$
Intraday, CV(%) [1]	2.41	3.33	3.11
Total precision, CV(%) [1]	2.46	3.35	3.53
Total accuracy, Relative Recovery (%) [2]	99.3	101.2	99.6

[1] Coefficient of variation; [2] Relative recovery percentage.

3.3. *Application to the Analysis of Real Samples*

The proposed method was applied to the analysis of three batches of Argireline peptide solution C® labelled to contain 0.05% *w/w* acetyl hexapeptide-8, and three batches of anti-wrinkle cosmetic cream labelled to contain 0.005% *w/w* of the peptide. Results obtained from the analysis of real cosmetic products are presented in Table 4.

Table 4. Quantitation of acetyl hexapeptide-8 in cosmetic formulation and cosmetic cream.

Cosmetic Product	% Label Claim (±SD) [1] ($n = 5$)	% CV [2]
Cosmetic formulation		
Batch No F1	0.0511 (±0.0015)	2.93
Batch No F2	0.0507 (±0.0011)	2.16
Batch No F3	0.0491 (±0.0016)	3.25
Cosmetic cream		
Batch No C1	0.00491 (±0.00018)	3.7
Batch No C2	0.00501 (±0.00013)	2.6
Batch No C3	0.00509 (±0.00021)	4.1

[1] Standard deviation; [2] Coefficient of variation.

The % recovery for the quantitation of acetyl hexapeptide-8 by the proposed HILIC-PDA method is ranged from 98.2 to 102.2 % in cosmetic formulation and from 98.2 to 101.8% in cosmetics creams.

4. Conclusions

There is a real need to set up analytical methods to quantitate the active compounds in cosmetic products incorporated with bioactive peptides in low content. In this work, a HILIC-PDA method was developed and validated for the determination of acetyl hexapeptide-8 in cosmetics. Over the past two decades the use of biopeptides in cosmetic products is increasingly expanded. The developed method takes full advantage of the benefits of HILIC leading to efficient retention of acetyl hexapeptide-8 with less matrix effect. Validation results demonstrate that the proposed method allows for the quantitation of acetyl hexapeptide-8 in both cosmetic formulations and cosmetics creams. The simplicity of sample preparation procedure and the short chromatographic run time of less than 10 min gives the method the capability for high sample throughput and it can be used to support quality control of cosmetic products containing low content of acetyl hexapeptide-8. There is no doubt that HILIC chromatography enables the determination of bioactive peptides in cosmetic products without the need for specialized detection methods. The proposed method could be extended for future applications in the analysis of various bioactive peptides used for cosmetic purposes.

Author Contributions: Conceptualization, I.P., A.V., E.T. and V.R.; methodology, V.R. and E.K.; validation, V.R., E.K. and I.P.; formal analysis, V.R. and E.K.; investigation, V.R., E.K., A.V. and I.P.; resources, I.P. and A.V.; writing-original draft preparation, E.K. and I.P.; writing-review and editing, I.P., A.V., E.T., E.K. and V.R.; supervision, I.P. and A.V. All authors have read and agreed to the published version of the manuscript.

Funding: This research received no external funding.

Institutional Review Board Statement: Not applicable.

Informed Consent Statement: Not applicable.

Data Availability Statement: Data are available upon request to the corresponding author.

Conflicts of Interest: The authors declare no conflict of interest.

References

1. Lupo, M.P.; Cole, A.L. Cosmeceutical peptides. *Derm. Ther.* **2007**, *20*, 343–349. [CrossRef]
2. Rivers, J.K. The role of cosmeceuticals in antiaging therapy. *Skin Ther. Lett.* **2008**, *13*, 5–9.
3. Zhang, L.; Falla, T.J. Cosmeceuticals and peptides. *Clin. Derm.* **2009**, *27*, 485–494. [CrossRef]
4. Gorouhi, F.; Maibach, H.I. Role of topical peptides in preventing or treating aged skin. *Int. J. Cosmet. Sci.* **2009**, *31*, 327–345. [CrossRef]
5. Husein el Hadmed, H.; Castillo, R.F. Cosmeceuticals: Peptides, proteins, and growth factors. *J. Cosmet. Derm.* **2016**, *15*, 514–519. [CrossRef]
6. Sutton, R.B.; Fasshauer, D.; Jahr, R.; Brunger, A.T. Crystal structure of a SNARE complex involved in synaptic exocytosis at 2.4. A resolution. *Nature* **1998**, *395*, 347–353. [CrossRef]

7. Blanes-Mira, C.; Clemente, J.; Jodas, G.; Gil, A.; Fernández-Ballester, G.; Ponsati, B.; Gutierrez, L.; Pérez-Payá, E.; Ferrer-Montiel, A. A synthetic hexapeptide (Argireline) with antiwrinkle activity. *Int. J. Cosmet. Sci.* **2002**, *24*, 303–310. [CrossRef]
8. Hoppel, M.; Reznicek, G.; Kahlig, H.; Kotisch, H.; Resch, G.P.; Valenta, C. Topical delivery of acetyl hexapeptide-8 from different emulsions: Influence of emulsion composition and internal structure. *Eur. J. Pharm. Sci.* **2015**, *68*, 27–35. [CrossRef]
9. Blanes-Mira, C.; Merino, J.M.; Valera, E.; Fernández-Ballester, G.; Gutiérrez, L.M.; Viniegra, S.; Pérez-Payá, E.; Ferrer-Montiel, A. Small peptides patterned after the N-terminus domain of SNAP25 inhibit SNARE complex assembly and regulated exocytosis. *J. Neurochem.* **2004**, *88*, 124–135. [CrossRef] [PubMed]
10. Grumelli, C.; Verderio, C.; Pozzi, D.; Rossetto, O.; Montecucco, C.; Matteoli, M. Internalization and mechanism of action of clostridial toxins in neurons. *Neurotoxicology* **2005**, *26*, 761–767. [CrossRef]
11. Mira, M.C.B.; Llobregat Hernandez, M.M.; Tebar, A.I.G.; Fernandez Ballester, G.J.; Planell Cases, R.M.; Ferrer Montiel, A.V.; Viniegra Bover, S.; Gutierrez Perez, L.M.; Carbonell Castello, T.; Perez Paya, E. Neuronal Exocytosis Inhibiting Peptides and Cosmetic and Pharmaceutical Compositions Containing Sad Peptides. U.S. Patent 7,015,192 B1, 21 March 2006. Available online: https://patentimages.storage.googleapis.com/dc/2f/ec/3e6fef57fd6254/US7015192.pdf (accessed on 29 July 2021).
12. Ruiz, M.A.; Clares, B.; Morales, M.E.; Cazalla, S.; Gallardo, V. Preparation and stability of cosmetic formulations with an anti-aging peptide. *J. Cosmet. Sci.* **2007**, *58*, 157–171. [CrossRef]
13. Jankovic, J.; Kenney, C.; Grafe, S.; Goertelmeyer, R.; Comes, G. Relationship between various clinical outcome assessments in patients with blepharospasm. *Mov. Disord.* **2009**, *24*, 407–413. [CrossRef]
14. Wang, Y.; Wang, M.; Xiao, X.S.; Huo, J.; Zhang, W.D. The anti-wrinkle efficacy of Argireline. *J. Cosmet. Laser Ther.* **2013**, *15*, 237–241. [CrossRef]
15. Tandini, K.A.; Mercurio, D.G.; Gonçalves Maia Campos, P.M.B. Acetyl hexapeptide-3 in a cosmetic formulation acts on skin mechanical properties-clinical study. *Braz. J. Pharm. Sci.* **2015**, *51*, 901–909. [CrossRef]
16. Raikou, V.; Varvaresou, A.; Panderi, I.; Papageorgiou, E. The efficacy study of the combination of tripeptide-10 citrulline and acetyl hexapeptide-3. A prospective randomized controlled study. *J. Cosmet. Dermatol.* **2017**, *16*, 271–278. [CrossRef]
17. Lungu, C.; Considine, E.; Zahir, S.; Ponsati, B.; Arrastiac, S.; Hallett, M. Pilot study of topical acetyl hexapeptide-8 in the treatment for blepharospasm in patients receiving botulinum toxin therapy. *Eur. J. Neurol.* **2013**, *20*, 515–518. [CrossRef]
18. Yoshida, T. Peptide separation by Hydrophilic-Interaction Chromatography: A review. *J. Biochem. Biophys. Meth.* **2004**, *60*, 265–280. [CrossRef]
19. Alpert, A.J. Hydrophilic-interaction chromatography for the separation of peptides, nucleic acids and other polar compounds. *J. Chromatogr. A* **1990**, *499*, 177–196. [CrossRef]
20. Kozlik, P.; Goldman, R.; Sanda, M. Hydrophilic interaction liquid chromatography in the separation of glycopeptides and their isomers. *Anal. Bioanal. Chem.* **2018**, *410*, 5001–5008. [CrossRef]
21. Roca, L.S.; Schoemaker, S.E.; Pirok, B.W.J.; Gargano, A.F.G.; Schoenmakers, P.J. Accurate modelling of the retention behaviour of peptides in gradient-elution hydrophilic interaction liquid chromatography. *J. Chromatogr. A* **2020**, *1614*, 460650. [CrossRef]
22. Molnarova, K.; Kozlík, P. Comparison of different HILIC stationary phases in the separation of hemopexin and immunoglobulin G glycopeptides and their isomers. *Molecules* **2020**, *25*, 4655. [CrossRef]
23. Hemström, P.; Irgum, K. Hydrophilic interaction chromatography. *J. Sep. Sci.* **2006**, *29*, 1784–1821. [CrossRef]
24. Fical, L.; Khalikova, M.; Kočová Vlčková, H.; Lhotská, I.; Hadysová, Z.; Vokřál, I.; Červený, L.; Švec, F.; Nováková, L. Determination of Antiviral Drugs and Their Metabolites Using Micro-Solid Phase Extraction and UHPLC-MS/MS in Reversed-Phase and Hydrophilic Interaction Chromatography Modes. *Molecules* **2021**, *26*, 2123. [CrossRef]
25. Panderi, I.; Malamos, Y.; Machairas, G.; Zaharaki, S. Investigation of the retention mechanism of cephalosporins by zwitterionic hydrophilic interaction liquid chromatography. *Chromatographia* **2016**, *79*, 995–1002. [CrossRef]
26. Panderi, I.; Taxiarchi, E.; Pistos, C.; Kalogria, E.; Vonaparti, A. Insights into the Mechanism of Separation of Bisphosphonates by Zwitterionic Hydrophilic Interaction Liquid Chromatography: Application to the Quantitation of Risedronate in Pharmaceuticals. *Separations* **2019**, *6*, 6. [CrossRef]
27. Dinh, N.P.; Jonsson, T.; Irgum, K. Water uptake on polar stationary phases under conditions for hydrophilic interaction chromatography and its relation to solute retention. *J. Chromatogr. A* **2013**, *1320*, 33–47. [CrossRef]
28. Jandera, P.; Hájek, T. Mobile phase effects on the retention on polar columns with special attention to the dual hydrophilic interaction–reversed-phase liquid chromatography mechanism, a review. *J. Sep. Sci.* **2018**, *41*, 145–162. [CrossRef] [PubMed]
29. Janviera, S.; De Sutter, E.; Wynendaele, E.; De Spiegeleer, B.; Vanhee, C.; Deconinck, E. Analysis of illegal peptide drugs via HILIC-DAD-MS. *Talanta* **2017**, *174*, 562–571. [CrossRef] [PubMed]
30. Stalmans, S.; Gevaert, B.; Verbeke, F.; D'Hondt, M.; Bracke, N.; Wynendaele, E.; De Spiegeleer, B. Quality control of cationic cell-penetrating peptides. *J. Pharm. Biomed. Anal.* **2016**, *117*, 289–297. [CrossRef] [PubMed]
31. Chirita, R.I.; Chaimbault, P.; Archambault, J.C.; Robert, I.; Elfakir, C. Development of a LC–MS/MS method to monitor palmitoyl peptides content in anti-wrinkle cosmetics. *Anal. Chim. Acta* **2009**, *641*, 95–100. [CrossRef]
32. Papagianni, P.; Varvaresou, A.; Papageorgiou, S.; Panderi, I. Development and validation of an ion-pair RP-HPLC method for the determination of oligopeptide-20 in cosmeceuticals. *J. Pharm. Biomed. Anal.* **2011**, *56*, 645–649. [CrossRef]
33. Giannakou, M.; Varvaresou, A.; Kiriazopoulos, E.; Papageorgiou, S.; Kavvalou, E.; Tsirivas, E.; Panderi, I. Quantification of oligopeptide-20 and oligopeptide-24 in cosmetic creams using hydrophilic interaction liquid chromatography with electrospray ionization mass spectrometry. *Sep. Sci. Plus* **2018**, *1*, 159–167. [CrossRef]

34. Zhou, W.; Wang, P.G.; Krynitsky, A.J.; Rader, J.I. Rapid and simultaneous determination of hexapeptides (Ac-EEMQRR-amide and H2N-EEMQRR-amide) in anti-wrinkle cosmetics by hydrophilic interaction liquid chromatography–solid phase extraction preparation and hydrophilic interaction liquid chromatography with tandem mass spectrometry. *J. Chromatogr. A* **2011**, *1218*, 7956–7963.
35. Kraeling, M.E.K.; Zhou, W.; Wang, P.; Ogunsola, O.A. In vitro skin penetration of acetyl hexapeptide- 8 from a cosmetic formulation. *J. Cutan. Ocular. Toxicol.* **2015**, *34*, 46–52. [CrossRef]
36. Zhou, W.; Wang, P.G.; Ogunsola, O.A.; Kraeling, M.E.K. Rapid determination of hexapeptides by hydrophilic interaction LC-MS/MS for in vitro skin penetration studies. *Bioanalysis* **2013**, *5*, 1353–1362. [CrossRef] [PubMed]
37. ICH. Guideline Q2(R1) Validation of Analytical Procedures: Text and Methodology. In Proceedings of the International Conference on Harmonisation, London, UK 2005. Available online: https://www.ema.europa.eu/en/documents/scientific-guideline/ich-q-2-r1-validation-analytical-procedures-text-methodology-step-5_en.pdf (accessed on 29 July 2021).
38. Machairas, G.; Panderi, I.; Geballa-Koukoula, A.; Rozou, S.; Antonopoulos, N.; Charitos, C.; Vonaparti, A. Development and validation of a hydrophilic interaction liquid chromatography method for the quantitation of impurities in fixed-dose combination tablets containing rosuvastatin and metformin. *Talanta* **2018**, *183*, 131–141. [CrossRef] [PubMed]
39. Johnsen, E.; Leknes, S.; Wilson, S.R.; Lundanes, E. Liquid chromatography-mass spectrometry platform for both small neurotransmitters and neuropeptides in blood, with automatic and robust solid phase extraction. *Sci. Rep.* **2015**, *5*, 1–16. [CrossRef] [PubMed]
40. McCalley, D.V. Understanding and manipulating the separation in hydrophilic interaction liquid chromatography. *J. Chromatogr. A* **2017**, *1523*, 49–71. [CrossRef] [PubMed]
41. Antonopoulos, N.; Machairas, G.; Migias, G.; Vonaparti, A.; Brakoulia, V.; Pistos, C.; Gennimata, D.; Panderi, I. Hydrophilic Interaction Liquid Chromatography-Electrospray Ionization Mass Spectrometry for Therapeutic Drug Monitoring of Metformin and Rosuvastatin in Human Plasma. *Molecules* **2018**, *23*, 1548. [CrossRef]
42. *British Pharmacopoeia*; Her Majesty's Stationery Office: London, UK, 2004; Volume II, pp. 1428–1429.

separations

MDPI

Article

Revisiting Chiral Recognition Mechanism on Chicken Alpha 1-Acid Glycoprotein: Location of Chiral Binding Sites and Insight into Chiral Binding Mechanism

Jun Haginaka [1,*], Taku Yamashita [2], Hirofumi Tsujino [3] and Mitsuhiro Arisawa [3]

[1] Institute for Biosciences, Mukogawa Women's University, 11-68, Koshien Kyuban-cho, Nishinomiya 663-8179, Japan
[2] School of Pharmacy and Pharmaceutical Sciences, Mukogawa Women's University, 11-68, Koshien Kyuban-cho, Nishinomiya 663-8179, Japan; taku@mukogawa-u.ac.jp
[3] Graduate School of Pharmaceutical Sciences, Osaka University, 1-6 Yamadaoka, Suita, Osaka 565-0871, Japan; htsujino@phs.osaka-u.ac.jp (H.T.); arisaw@phs.osaka-u.ac.jp (M.A.)
* Correspondence: haginaka@mukogawa-u.ac.jp

Abstract: Chiral stationary phases based on chicken alpha 1-acid glycoprotein (cAGP) have been used for enantioseparations of various compounds. However, the chiral binding sites and mechanism have not been clarified yet. Based on chromatographic properties of native and W26-modified cAGP columns and docking simulations of studied compounds into the generated model structure of cAGP, the chiral binding sites were located on cAGP and the chiral binding mechanism was discussed. On cAGP, there existed a binding cavity lined with H25, W26, Y47, R128, T129, D161 and E168, which contribute electrostatic or hydrogen bonding interactions. Benzoin and chlorpheniramine enantiomers interacted with cAGP at almost the same sites a little away from W26, while propranolol enantiomers docked, slightly shifting toward H25 and W26. Furthermore, in addition to hydrophobic interactions, ionic interactions between amino groups of chlorpheniramine enantiomers and a carboxyl group of D161 or E168 played an important role in the chiral recognition, while hydrophobic interactions and hydrogen bonding interactions worked for the chiral recognition of benzoin and propranolol enantiomers.

Keywords: protein-based chiral stationary phase; alpha 1-acid glycoprotein; chiral recognition mechanism; molecular docking

Citation: Haginaka, J.; Yamashita, T.; Tsujino, H.; Arisawa, M. Revisiting Chiral Recognition Mechanism on Chicken Alpha 1-Acid Glycoprotein: Location of Chiral Binding Sites and Insight into Chiral Binding Mechanism. *Separations* **2021**, *8*, 73. https://doi.org/10.3390/separations8060073

Academic Editor: Victoria Samanidou

Received: 6 April 2021
Accepted: 19 May 2021
Published: 23 May 2021

1. Introduction

Chiral stationary phases (CSPs) based on proteins have been used for enantioseparations of racemates or chiral compounds [1,2]. Those include serum proteins such as bovine serum albumin [3], human serum albumin [4], human alpha 1-acid glycoprotein (hAGP) [5] and chicken AGP (cAGP) [6,7]; enzymes such as trypsin [8], cellobiohydrolase [9], pepsin [10] and amyloglucosidase [11]; and other proteins such as antibody [12] and fatty acid binding protein [13]. The disadvantages of protein-based CSPs include low capacity, lack of column ruggedness and limited understanding of the chiral recognition mechanism. The advantages are that various drug enantiomers could be separated in reversed-phase mode because of multiple binding interactions such as hydrophobic, hydrogen bonding and electrostatic interactions or multiple binding sites [1]. Among those CSPs, a CSP based on hAGP could separate a wide range of basic, neutral and acidic enantiomers [1]. hAGP consisted of 183 amino acid residues, which contain five *N*-linked glycans, and accounted for the total 37–54 kDa mass [14,15]. It belongs to a member of the lipocalin family, which is a transporter for small hydrophobic molecules [16,17].

Furthermore, hAGP also binds to a variety of drugs, mainly hydrophobic or basic drugs because of its isoelectric point (pI) 2.7–3.8 [16]. Besides the high heterogeneity of glycans, the protein part has also been found to show polymorphism [18]. The variants are encoded by two different genes: The F1*S variant is encoded by the alleles of the same gene (orosomucoid1 (ORM1)), while the A variant is encoded by a different gene (ORM2). The molar ratio of the F1*S and A variant hAGPs in blood typically ranges from 3:1 to 2:1 [19,20]. The first X-ray structural analysis of the F1*S variant hAGP was reported, and its ligand-binding sites, which consist of the lobes I, II and III, were clarified. Lobe I is a large and deep cavity, and provides sufficient spaces for ligand-binding pockets. Lobes II and III are on each side of lobe I and are negatively charged. Furthermore, the x-ray crystal structure of the A variant hAGP, which had only two lobes, lobes I and II, and had narrower binding sites than the F1*S variant hAGP, were determined [21]. The drug bindings to hAGP have been extensively investigated using co-crystals with ligands [22], docking study with the 3D molecular model [23–30], circular dichroism (CD) and fluorescence spectroscopy [23,31,32], photoaffinity labelling [33] and NMR spectroscopy [34]. Although most of studies indicated that the ligand-binding sites were located on lobes I–III, only one study by induced CD spectra suggested that W25 of hAGP was essentially involved in the ligand-bindings [32].

Regarding chiral binding mechanism of ligands on cAGP, a few studies were reported [24,34]. Chiral bindings of the coumarins (warfarin, acenocoumarol and phenprocoumon) to hAGPs of the F1*S and A variants (ORM1 and ORM2, respectively) were examined [24]. The enantiomers of the coumarins were docked onto the F1*S and A variant hAGPs in essentially the same position, including W122, and orientation [24]. (*S*)-Enantiomers of warfarin and acenocoumarol were found to possess a higher affinity for ORM 1 than the corresponding (*R*)-enantiomers in binding measurements by stereoselective analysis of the ultrafiltrates [24]. (*R*)- and (*S*)-acenocoumarol to hAGP models of the F1*S and A variants show a slight preference for (*S*)-enantiomer in the case of both variant hAGPs [24]. However, the proposed model could not explain chiral bindings of coumarins onto hAGP. Chiral bindings of propranolol enantiomers to cAGP were investigated with ligand-detected NMR experiments [34]. It was found that each enantiomer bound to the AGP binding pocket in a different orientation.

CSPs based on crude ovomucoid, ovomucoid from chicken (OMCHI), which was isolated from egg whites, were developed [35] and used for enantioseparations of neutral, acidic and basic compounds [36]. However, crude ovomucoid included OMCHI and ovoglycoprotein (OGCHI), whose content was about 10%. Since OMCHI and OGCHI had the same pI 4.1 [1], OMCHI was contaminated with OGCHI in crude ovomucoid preparation [6]. Furthermore, OMCHI had no chiral recognition ability, but the chiral recognition ability of crude ovomucoid came from OGCHI [6]. The amino acid sequence of OGCHI was clarified [7]. OGCHI consisted of 183 amino acids and accounted for the total 30 kDa mass. Furthermore, OGCHI showed 31–32% identities to rabbit AGP and hAGP [7]. OGCHI was a member of the AGP family identified in mammals, and was chicken AGP (cAGP) [7]. As hAGP, cAGP conserved the number and locations of the S–S bridges (C6-146 and C73–C163), and had five *N*-linked glycans: N16, 70, 77 and 87 were fully glycosylated and N62 was partially glycosylated [37].

cAGP had only one tryptophan residue at 26 position [37]. W26-modified cAGP completely lost chiral recognition ability for β-blockers such as propranolol, alprenolol and oxprenolol, while it still conserved lower chiral recognition ability for chlorpheniramine and benzoin, compared with native cAGP [38]. Therefore, it was concluded that β-blockers bound on a single binding site near the W26 region and that further bindings of chlorpheniramine and benzoin occurred at the secondary binding site [38]. Furthermore, by CD and ultraviolet/visible absorption spectroscopy experiments it was supported that W26 was responsible for the ligand-bindings to cAGP [39].

In this study, we located the chiral binding sites on cAGP and discussed the chiral recognition mechanism of benzoin, chlorpheniramine and propranolol, based on chromatographic properties of native and W26-modified cAGP columns and molecular docking results.

2. Materials and Methods

2.1. Reagents

2-Nitrophenylsulfenyl chloride was purchased from Tokyo Chemical Industry (Tokyo, Japan). Other reagents and solvents were obtained from Nacalai Tesque (Kyoto, Japan). All reagents were of an analytical-reagent grade and were used without further purification. The structures of studied compounds were illustrated in Figure 1.

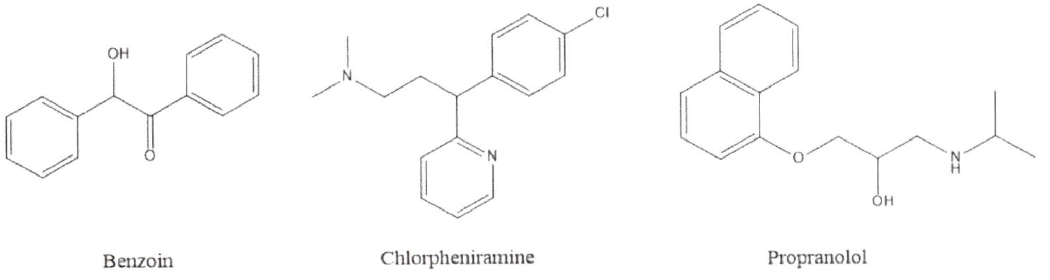

Figure 1. Structures of compounds used in this study.

2.2. Preparation of W26-Modified cAGP

cAGP was isolated as reported previously [6]. W26-modified cAGP with 2-nitrophenylsulfenyl chloride was obtained according to a previously reported method [40]. Briefly, 3 mg of 2-nitrophenylsulfenyl chloride dissolved in 50 μL of glacial acetic acid were added to a 3 mL of a 20 vol% acetic acid solution including 100 mg of cAGP. Then the reaction mixture was stirred for 5 h at 0 °C. The reaction mixture was dialyzed against water for 60 h at 4 °C and lyophilized.

2.3. Preparation of Native and W26-Modified cAGP Columns

Native and W26-modified cAGP columns were prepared as reported previously [6]. Briefly, 2.66 μmol of native and W26-modified cAGPs, respectively, were reacted with 1 g of aminopropyl-silica gels (5 μm particle size, 12 nm pore size) activated with *N,N*-disuccinimidyl carbonate (DSC). Each protein reacted was completely immobilized to DSC-activated aminopropyl-silica gels [41]. The obtained materials were packed into a 2.0 mm I.D. × 100 mm stainless-steel column by a slurry packing method to evaluate chiral recognition abilities. The flow-rate was maintained at 0.2 mL/min. Detection was carried out at 210 nm. The retention factor, k, and enantioseparation factor, α, of a racemate were calculated according to a method reported previously [6]. All separations were performed at 25 °C using a water bath.

2.4. Generation of Model Structure of cAGP and Docking Simulations

The homology model of cAGP was constructed using Swiss-Model server [42]. The crystal structure of hAGP (A variant (ORM2); pdbid, 3apu), which showed the highest score of 28.8% homology with cAGP in amino acids sequence alignments, was selected, and the model structure was built in accordance with the alignment with the target protein using ProMod3 [43]. In addition, the model structure of cAGP was generated using i-TASSER server [44], and the built structure seemed to be indistinguishable with that from Swiss-Model server. The validated model structure from Swiss-Model server was used for docking simulations as previously reported [45–47]. First, the model structure was prepared by Protein Preparation Wizard, which provides a state for calculation with charge, orientation of groups, etc. Next, the chemical structures for benzoin, chlorpheniramine and propranolol were obtained from Pubchem database (https://pubchem.ncbi.nlm.nih.gov/, accessed on 22 May 2021) for both (*R*)- and (*S*)-configurations, and feasible conformations were prepared by LigPrep module. Each prepared ligand was docked to the built model structure of cAGP using Glide module using standard precision (SP) mode [48], and the docked structures with the highest glide score for each ligand were visualized with measuring distances among atoms by the PyMOL Molecular Graphics System from Schrödinger (New York, NY, USA). All tools for docking and visualization were from Schrödinger.

3. Results

3.1. Enantioseparations on Native and W26-Modified cAGP Columns

Since cAGP had only one tryptophan residue at 26 position, W26 of cAGP was only modified with 2-nitrophenylsulfenyl chloride. Table 1 shows retention and enantioseparation factors of benzoin, chlorpheniramine and propranolol on native and W26-modified cAGP columns. Three racemic compounds were enantioseparated on a native cAGP column. Though on a W26-modified cAGP column, retention and enantioseparation factors of benzoin and chlorpheniramine were decreased, they were still enantioseparated. However, propranolol was not enantioseparated on a W26-modified cAGP column along with a drastic decrease in the retention factor. Similarly, chiral resolution of β-blockers such as alprenolol and pindolol was completely lost on a W26-modified cAGP column [38]. Furthermore, competition studies using *N,N*-dimethyl-*n*-octylamine (DMOA) as a competitor indicated that propranolol competed with DMOA on a single binding site near the W26 region and that further bindings of benzoin and chlorpheniramine occurred at the secondary binding site in a non-competitive fashion with DMOA [38]. Therefore, it was considered that there existed at least two chiral binding sites on native cAGP.

Table 1. Comparison of chiral recognition abilities of native and W26-modified cAGP columns [1].

Entry	Native cAGP Column			W26-modified cAGP Column		
	k_S [2]	k_R [2]	α [3]	k_S [2]	k_R [2]	α [3]
Benzoin	5.97	19.5	3.27	4.82	8.00	1.66
Chlorpheniramine	8.40	4.49	1.87	4.03	2.76	1.46
Propranolol	27.0	30.5	1.13	2.06	2.06	1.00

[1] Data adapted from Table 1 in [38]. [2] Retention factors of (*S*)- and (*R*)-enantiomers, k_S and k_R, respectively. [3] Enantioseparation factor of (*S*)- and (*R*)-enantiomers.

3.2. Docking Simulations of Studied Compounds into Generated Model Structure of cAGP

As described above, the model structure of cAGP was generated using hAGP as a template, and then the generated structure was validated with another modeling server. Subsequently, each enantiomer of three compounds, benzoin, chlorpheniramine and propranolol, was simulated and docked to a certain cavity on the generated model structure of cAGP (Figure 2).

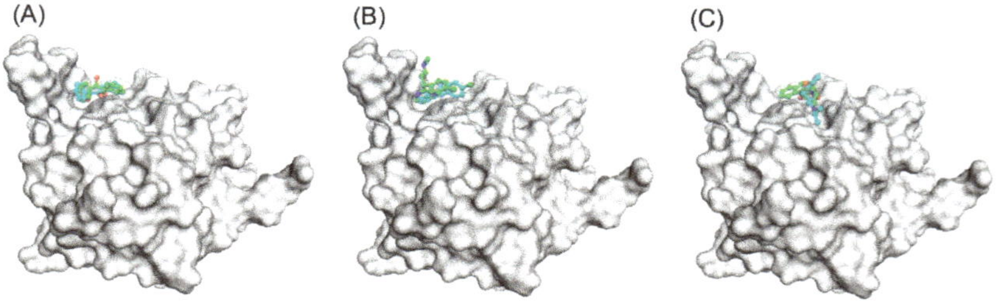

Figure 2. Docking structures of cAGP with (**A**) benzoin, (**B**) chlorpheniramine and (**C**) propranolol enantiomers. The built model structure of cAGP was illustrated as surface mode and the docked enantiomers were shown as ball-and-stick models, respectively.

Remarkably, each enantiomer of two compounds, benzoin and chlorpheniramine, was located onto a similar position of the surface; however the docked propranolol enantiomers slightly shifted to the relatively small cliff. The differences between (*R*)- and (*S*)-configurations in the binding site were subsequently investigated in detail.

As shown in Figure 3, both (*R*)- and (*S*)-benzoin were docked onto a similar position of a cavity of the generated model structure of cAGP. The positions seemed to be fastened by one of aromatic rings in both configurations by hydrophobic interactions. Furthermore, some of hydrogen bonding interactions work for chiral recognition of (*R*)- and (*S*)-benzoin. A hydroxyl group of (*R*)-benzoin is close to oxygen atoms in either a carbonyl or hydroxyl group of D161 with a distance of 3.2 or 4.9 Å, and a hydroxyl group of Y47 is nearby a carbonyl group of (*R*)-benzoin with a distance of 2.8 Å (Figure 3A). An oxygen atom in either a carbonyl or hydroxyl group of D161 is also close to a hydroxyl group of (*S*)-benzoin at a distance of 2.8 Å; however, Y47 is a bit far from the bound (*S*)-benzoin at a distance of 5.3 Å (Figure 3B). These results indicate that (*R*)-benzoin binds to cAGP more tightly than (*S*)-benzoin.

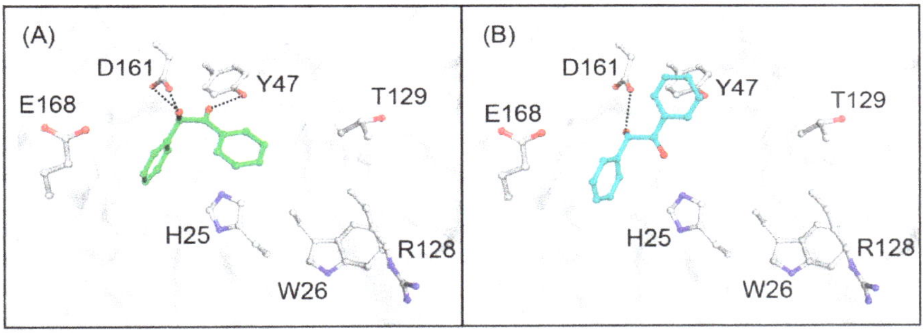

Figure 3. Simulated docking structures of cAGP focusing on the surrounding of bound (**A**) (*R*)-benzoin and (**B**) (*S*)-benzoin. The docked compounds were represented in either green or cyan for carbon atoms. Besides, nitrogen and oxygen atoms were colored in blue and red, respectively. The remarked amino acid residues were labeled and shown in gray for carbon atoms, and feasible interactions were depicted as dotted lines.

Next, (*R*)- and (*S*)-chlorpheniramine were also suggested to bind the same cavity with benzoin (Figure 4). As shown in Figure 4, a carboxyl group of E168 is close to an amino group of (*R*)-chlorpheniramine at a distance of 2.6 Å, while an amino group of (*S*)-chlorpheniramine could interact with a carboxyl group of D161 with 3.0 Å instead of E168 for (*R*)-chlorpheniramine. Electrostatic interactions between amino groups of (*R*)- and (*S*)-chlorpheniramine and carboxyl groups of E168 and D161, respectively, could work for the recognition. In addition, as shown in Figure 4B, Y47 was located nearby bound (*S*)-chlorpheniramine with a distance of 3.9 Å. Furthermore, the chlorine atom of (*S*)-chlorpheniramine was close to a carbonyl group for main chain between R128 and T129 with a distance of 3.6 Å. It implied that (*S*)-chlorpheniramine could interact with the main chain via halogen bonding interactions. These results indicate that (*S*)-chlorpheniramine interacts with cAGP more tightly than (*R*)-chlorpheniramine.

Finally, (*R*)- and (*S*)-propranolol were again docked to the generated model structure of cAGP (Figure 5). Interestingly, the docked propranolol was slightly shifted toward H25 and W26 with a consequent distance of 4.3 Å to H25. Although Y47 was again suggested to be involved in the binding of (*R*)-propranolol with a distance of 4.7 Å, it could be weak to anchor the docked propranolol. A carbonyl group for main chain between R128 and T129 was remarkably located to either 2.7 or 2.8 Å for (*R*)-propranolol or (*S*)-propranolol. The former hydrogen bonding interactions were with an amino group of (*R*)-propranolol, while the latter hydrogen bonding interactions were with a hydroxyl group of (*S*)-propranolol.

These results indicate that the (*R*)-propranolol binding to cAGP is more favorable than the (*S*)-propranolol binding.

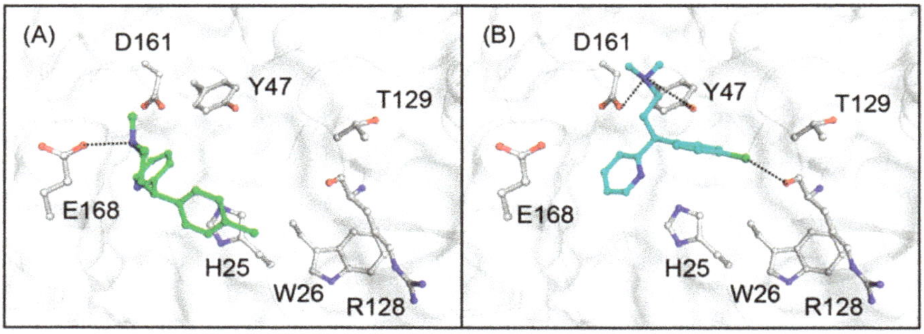

Figure 4. Simulated docking structures of cAGP with (**A**) (*R*)-chlorpheniramine and (**B**) (*S*)-chlorpheniramine. The docked compounds were represented as in Figure 3. Besides, the same colors were used for nitrogen and oxygen atoms, and the remarked amino acid residues as in Figure 3. Feasible interactions were depicted as dotted lines.

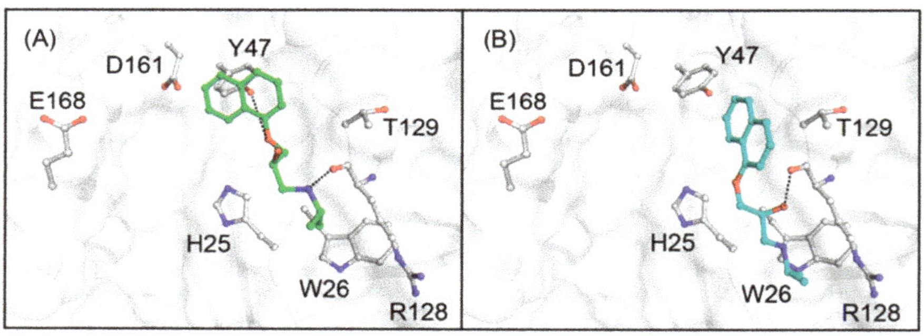

Figure 5. Simulated docking structures of cAGP with (**A**) (*R*)-propranolol and (**B**) (*S*)-propranolol. The docked compounds were represented as in Figure 3. Besides, the same colors were used for nitrogen and oxygen atoms, and the remarked amino acid residues as in Figure 3. Feasible interactions were depicted as dotted lines.

4. Discussion

Figure 6 shows sequence alignments of F1*S and A variant hAGPs and cAGP along with the highlighted amino acids, which consist of feasible ligand-binding sites. The F1*S variant hAGP has lobs I, II and III as the ligand-binding sites, while the A variant hAGP has only lobs I and II. Figure 7 shows crystal structures of F1*S (left; pdbid, 3kq0) and A (center; pdbid, 3apu) variant hAGPs and a built model structure of cAGP (right). These indicate that the ligand-binding sites of hAGP and cAGP are totally different: A lot of ligands bound almost the same sites on hAGP, lobes I-III, while the three ligands, benzoin, chlorpheniramine and propranolol, bound the cavity near a W26 region on cAGP. The hydrophobic cavities lined with H25, W26, Y47, R128, T129, D161 and E168, which contribute electrostatic or hydrogen bonding interactions, were responsible for chiral resolution of the studied compounds on cAGP.

```
F1*S hAGP   -QIPLCANLV PVPITNATLD QITGKWFYIA SAFRNEEYNK SVQEIQATFF
A hAGP      MQIPLCANLV PVPITNATLD RITGKWFYIA SAFRNEEYNK SVQEIQATFF
cAGP        TESPTCAPLV TADMDNATVD RLLGHWVYIM GASQYPPHMA EMRELKYATF

F1*S hAGP   YFTPNKTEDT IFLREYQTRQ DQCIYN-TTY LNVQRENGTI SRYVGGQEHF
A hAGP      YFTPNKTEDT IFLREYQTRQ NQCFYN-SSY LNVQRENGTV SRYEGGREHV
cAGP        TLFPGSHEDE FNVTEIMRLN ETCVVKNSSK IHVFRHNSTL THEDG---QV

F1*S hAGP   AHLLILRDTK TYMLAFDVND EKNWGLSVYA DKPETTKEQL GEFYEALDCL
A hAGP      AHLLFLRDTK TLMFGSYLDD EKNWGLSFYA DKPETTKEQL GEFYEALDCL
Chicken     VSMAELIHSD KDLFILKHFK DNHVGLSLSA RTAEVTKEQL EEFEAQLRCH

F1*S hAGP   RIPKSDVVYT DWKKDKCEPL EKQHEKERKQ EEGESAWSHPQFEK
A hAGP      RIPRSDVMYT DWKKDKCEPL EKQHEKERKQ EEGESHHHHHH
cAGP        GFKLEEAFIT SPKDACPAAG EETGEGSAAP AEPQLG
```

Figure 6. Sequence alignments of F1*S and A variant hAGPs and cAGP. The cavities in hAGP were called as lobes being highlighted in green for lobe I, orange for lobe II and cyan for lobe III. The amino acids, which consisted of feasible ligand-binding sites in this work, were colored in magenta.

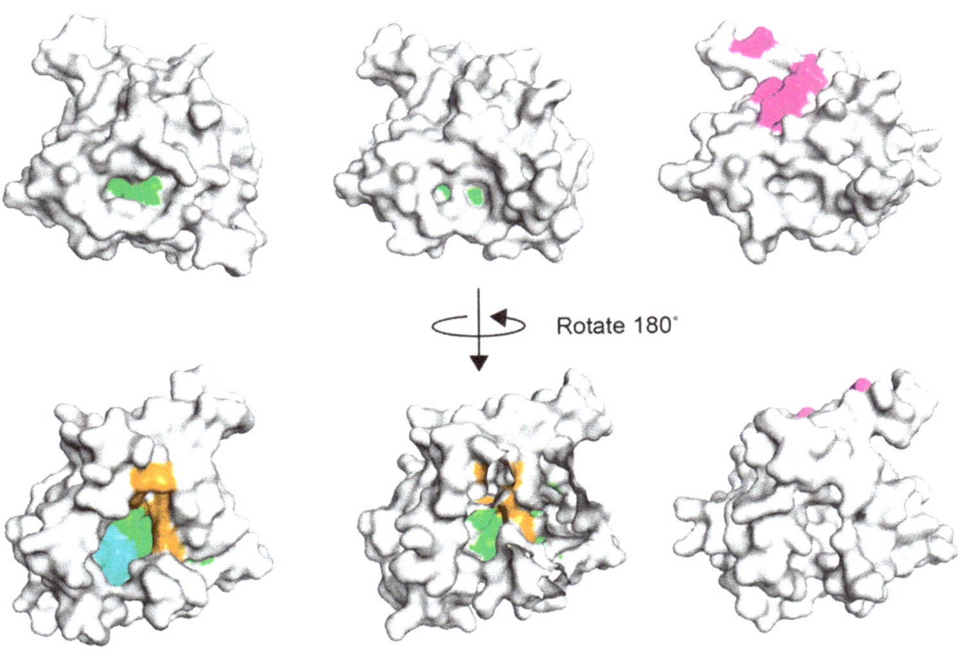

Figure 7. Crystal structures of F1*S (**left**) and A (**middle**) variant hAGPs and a built model structure of cAGP (**right**). The A variant of cAGP was only showed one of subunits. The highlighted residues in the Figure 6 were colored in the structures.

In docking simulations of studied enantiomers into the generated model structure of cAGP, the (*R*)-benzoin, (*S*)-chlorpheniramine or (*R*)-propranolol binding was more favorable than its enantiomer. These results correlate well with the chromatographic results that (*R*)-benzoin, (*S*)-chlorpheniramine or (*R*)-propranolol is more retained than its enantiomer on a cAGP column. Furthermore, chiral resolution of propranolol was completely lost on a W26-modified cAGP column. This could be due to the small changes in the position of the side chains of amino acid residues in chiral binding sites near W26. However, on a W26-modified cAGP column chiral resolution of benzoin and chlorpheniramine was still attained. Because their chiral binding sites were a little away from W26. Previously, we considered that there existed at least two chiral binding sites on cAGP. Taking into account the results of docking simulations, we conclude that the chiral binding sites for the studied compounds are near a W26 region.

In docking simulations of chlorpheniramine enantiomers, electrostatic interactions between amino groups of (*R*)- and (*S*)-chlorpheniramine with carboxyl groups of E168 and D161 of cAGP, respectively, worked for chiral recognition. It was reported that the retention and enantioseparation factors increased with an increase in mobile-phase pH on a crude ovomucoid (contaminated with cAGP) column [36]. This suggests that the ionic interactions work for the retention and chiral recognition of chlorpheniramine enantiomers. On the other hand, in docking simulations of propranolol enantiomers, plausible ionic-interactions between an amino group of each propranolol enantiomer with a carboxyl group of cAGP were not found. Consequently, the results of docking simulations well coincide with the data that the retention factor increases with an increase in mobile-phase pH on a crude ovomucoid (contaminated with cAGP) column, but that the enantioseparation factor remains unchanged [49]. The retention of propranolol was governed by nonspecific interactions of propranolol with cAGP, while the enantioselectivity came from the specific interactions of propranolol in the chiral binding site near W26. We also constructed the model structure of W26-modified cAGP with 2-ntirophenylsulfenyl group (data not shown). Although the generated model structure was less reliable, it seemed that the introduced 2-ntirophenylsulfenyl group could yield enhanced rigidity for the cAGP structure, and that small changes in the position of the side chains of amino acid residues occurred. These changes might result in the loss of enantioselectivity of propranolol on the W26-modified column.

5. Conclusions

Based on chromatographic properties of native and W26-modified cAGP columns and docking simulations of studied compounds into the generated model structure of cAGP, the chiral binding sites were located on cAGP for the first time and the chiral binding mechanism was discussed. On cAGP, there existed a hydrophobic cavity lined with H25, W26, Y47, R128, T129, D161 and E168. Benzoin and chlorpheniramine enantiomers interacted with cAGP at almost the same sites a little away from W26, while propranolol enantiomers docked in the relatively small cliff, slightly shifting toward H25 and W26. Furthermore, in addition to hydrophobic interactions, ionic interactions between amino groups of chlorpheniramine enantiomers and a carboxy group of D161 or E168 played an important role in the chiral recognition, while hydrophobic interactions and hydrogen bonding interactions worked for the chiral recognition of benzoin and propranolol enantiomers.

Author Contributions: Conceptualization, J.H. and T.Y.; methodology, J.H, T.Y., H.T. and M.A.; formal analysis, J.H, T.Y., H.T. and M.A.; investigation, J.H, T.Y., H.T. and M.A.; resources, J.H.; data curation, J.H, T.Y., H.T. and M.A.; writing—original draft preparation, J.H.; writing—review and editing, J.H, T.Y., H.T. and M.A.; visualization, T.Y., H.T. and M.A.; supervision, J.H.; project administration, J.H. and T.Y.; funding acquisition, J.H. All authors have read and agreed to the published version of the manuscript.

Funding: This work was partially supported by a funding from Daicel (Tokyo, Japan).

Data Availability Statement: Not applicable.

Acknowledgments: Not applicable.

Conflicts of Interest: The authors declare no conflict of interest.

References

1. Haginaka, J. Mechanistic Aspects of Chiral Recognition on Protein-Based Stationary Phases. *Adv. Chromatogr.* **2011**, *49*, 37–69. [CrossRef]
2. Bi, C.; Zheng, X.; Azaria, S.; Beeram, S.; Li, Z.; Hage, D.S. Chromatographic Studies of Protein-Based Chiral Separations. *Separations* **2016**, *3*, 27. [CrossRef] [PubMed]
3. Allenmark, S.; Bomgren, B.; Borén, H. Direct Liquid Chromatographic Separation of Enantiomers on Immobilized Protein Stationary Phases. IV. Molecular Interaction Forces and Retention Behaviour in Chromatography on Bovine Serum Albumin as a Stationary Phase. *J. Chromatogr.* **1984**, *316*, 617–624. [CrossRef]
4. Domenici, E.; Bertucci, C.; Salvadori, P.; Felix, G.; Cahagne, I.; Motellier, S.; Wainer, I.W. Synthesis and Chromatographic Properties of an HPLC Chiral Stationary Phase Based upon Human Serum Albumin. *Chromatographia* **1990**, *29*, 170–176. [CrossRef]
5. Hermansson, J. Direct Liquid Chromatographic Resolution of Racemic Drugs Using α1-Acid Glycoprotein as the Chiral Stationary Phase. *J. Chromatogr. A* **1983**, *269*, 71–80. [CrossRef]
6. Haginaka, J.; Seyama, C.; Kanasugi, N. The Absence of Chiral Recognition Ability in Ovomucoid: Ovoglycoprotein-bonded HPLC Stationary Phases for Chiral Recognition. *Anal. Chem.* **1995**, *67*, 2539–2547. [CrossRef]
7. Sadakane, Y.; Matsunaga, H.; Nakagomi, K.; Hatanaka, Y.; Haginaka, J. Protein Domain of Chicken α1-Acid Glycoprotein is Responsible for Chiral Recognition Ability. *Biochem. Biophys. Res. Commun.* **2002**, *295*, 587–590. [CrossRef]
8. Thelohan, S.; Jadaud, P.; Wainer, I.W. Immobilized Enzymes as Chromatographic Phases for HPLC: The Chromatography of Free and Derivatized Amino Acids on Immobilized Trypsin. *Chromatographia* **1989**, *28*, 551–555. [CrossRef]
9. Erlandsson, P.; Marle, I.; Hansson, L.; Isaksson, R.; Petterson, C.; Petterson, G. Immobilized Cellulase (CBH I) as a Chiral Stationary Phase for Direct Resolution of Enantiomers. *J. Am. Chem. Soc.* **1990**, *112*, 4573–4574. [CrossRef]
10. Haginaka, J.; Miyano, Y.; Saizen, Y.; Seyama, C.; Murashima, T. Separation of Enantiomers on a Pepsin-bonded Column. *J. Chromatogr. A* **1995**, *708*, 161–168. [CrossRef]
11. Nystrom, A.; Strandberg, A.; Aspergren, A.; Behr, S.; Karlsson, A. Use of Immobilized Amyloglucosidase as Chiral Selector in Chromatography. Immobilization and Performance in Liquid Chromatography. *Chromatographia* **1999**, *50*, 209–214. [CrossRef]
12. Hofstetter, H.; Hofstetter, O. Antibodies as Tailor-made Chiral Selectors for Detection and Separation of Stereoisomers. *Trends Anal. Chem.* **2005**, *24*, 869–879. [CrossRef]
13. Massolini, G.; de Lorenzi, E.; Calleri, E.; Bertucci, C.; Monaco, H.L.; Perduca, M.; Caccialanza, G.; Wainer, I.W. Properties of a Stationary Phase Based on Immobilized Chicken Liver Basic Fatty Acid-binding Protein. *J. Chromatogr. B* **2001**, *751*, 117–130. [CrossRef]
14. Schmid, K.; Nimerg, R.B.; Kimura, A.; Yamaguchi, H.; Binette, J.P. The Carbohydrate Units of Human Plasma Alpha1-acid Glycoprotein. *Biochim. Biophys. Acta* **1977**, *492*, 291–302. [CrossRef]
15. Luo, Z.; Lei, H.; Sun, Y.; Liu, X.; Su, D.F. Orosomucoid, an Acute Response Protein with Multiple Modulating Activities. *J. Physiol. Biochem.* **2015**, *71*, 329–340. [CrossRef]
16. Kremer, J.M.; Wilting, J.; Janssen, L.H. Drug Binding to Human Alpha-1-acid Glycoprotein in Health and Disease. *Pharmacol. Rev.* **1988**, *40*, 1–47.
17. Israili, Z.H.; Dayton, P.G. Human Alpha-1-Glycoprotein and Its Interactions with Drugs. *Drug Metab. Rev.* **2001**, *33*, 161–235. [CrossRef]
18. Dente, L.; Pizza, M.G.; Metspalu, A.; Cortese, R. Structure and Expression of the Genes Coding for Human Alpha 1-acid Glycoprotein. *EMBO J.* **1987**, *6*, 2289–2296. [CrossRef] [PubMed]
19. Eap, C.B.; Baumann, P. The Genetic Polymorphism of Human Alpha 1-acid Glycoprotein. *Prog. Clin. Biol. Res.* **1989**, *300*, 111–125.
20. Yuasa, I.; Weidinger, S.; Umetsu, K.; Suenaga, K.; Ishimoto, G.; Eap, B.C.; Duche, J.C.; Baumann, P. Orosomucoid System: 17 Additional Orosomucoid Variants and Proposal for a New Nomenclature. *Vox Sang.* **1993**, *64*, 47–55. [CrossRef]
21. Schönfeld, D.L.; Ravelli, R.B.; Mueller, U.; Skerra, A. The 1.8-Å Crystal Structure of Alpha1-acid Glycoprotein (Orosomucoid) Solved by UV RIP Reveals the Broad Drug-binding Activity of This Human Plasma Lipocalin. *J. Mol. Biol.* **2008**, *384*, 393–405. [CrossRef]
22. Nishi, K.; Ono, T.; Nakamura, T.; Fukunaga, N.; Izumi, M.; Watanabe, H.; Suenaga, A.; Maruyama, T.; Yamagata, Y.; Curry, S.; et al. Structural Insights into Differences in Drug-binding Selectivity between Two Forms of Human Alpha1-acid Glycoprotein Genetic Variants, the A and F1*S Forms. *J. Biol. Chem.* **2011**, *286*, 14427–14434. [CrossRef]
23. Zsila, F.; Bikádi, Z.; Simonyi, M. Induced Circular Dichroism Spectra Reveal Binding of the Antiinflammatory Curcumin to Human Alpha1-acid Glycoprotein. *Bioorg. Med. Chem.* **2004**, *12*, 3239–3245. [CrossRef]
24. Hazai, E.; Visy, J.; Fitos, I.; Bikádi, Z.; Simonyi, M. Selective Binding of Coumarin Enantiomers to Human Apha1-acid Glycoprotein Genetic Variants. *Bioorg. Med. Chem.* **2006**, *14*, 1959–1965. [CrossRef] [PubMed]
25. Azad, M.A.; Huang, J.X.; Cooper, M.A.; Roberts, K.D.; Thompson, P.E.; Nation, R.L.; Li, J.; Velkov, T. Structure-activity Relationships for the Binding of Polymyxins with Human α-1-Acid Glycoprotein. *Biochem Pharmacol.* **2012**, *84*, 278–291. [CrossRef]

26. Huang, J.X.; Cooper, M.A.; Baker, M.A.; Azad, M.A.; Nation, R.L.; Li, J.; Velkov, T. Drug-binding Energetics of Human α-1-Acid Glycoprotein Assessed by Isothermal Titration Calorimetry and Molecular Docking Simulations. *J. Mol. Recognit.* **2012**, *25*, 642–656. [CrossRef] [PubMed]
27. Ajmal, M.R.; Abdelhameed, A.S.; Alam, P.; Khan, R.H. Interaction of New Kinase Inhibitors Cabozantinib and Tofacitinib with Human Serum Alpha-1 Acid Glycoprotein. A Comprehensive Spectroscopic and Molecular Docking Approach. *Spectrochim. Acta A Mol. Biomol. Spectrosc.* **2016**, *159*, 199–208. [CrossRef]
28. Ajmal, M.R.; Almutairi, F.; Zaidi, N.; Alam, P.; Siddiqi, M.K.; Khan, M.V.; Zaman, M.; Ishtikhar, M.; Khan, R.H. Biophysical Insights into the Interaction of Clofazimine with Human Alpha 1-acid Glycoprotein: A Multitechnique Approach. *J. Biomol. Struct. Dyn.* **2019**, *37*, 1390–1401. [CrossRef] [PubMed]
29. Nishi, K.; Sakurama, K.; Kobashigawa, Y.; Morioka, H.; Udo, N.; Hashimoto, M.; Imoto, S.; Yamasaki, K.; Otagiri, M. Interaction of Aripiprazole With Human α_1-Acid Glycoprotein. *J. Pharm. Sci.* **2019**, *108*, 3911–3916. [CrossRef]
30. Wang, B.L.; Kou, S.B.; Lin, Z.Y.; Shi, J.H.; Liu, Y.X. Insights on the Interaction Mechanism of Brigatinib to Human α-1-Acid Glycoprotein: Experimental and Computational Approaches. *Int. J. Biolog. Macromolec.* **2020**, *157*, 340–349. [CrossRef] [PubMed]
31. Miyoshi, T.; Sukimoto, K.; Otagiri, M. Investigation of the Interaction Mode of Phenothiazine Neuroleptics with Apha 1-acid Glycoprotein. *J. Pharm. Pharmacol.* **1992**, *44*, 28–33. [CrossRef] [PubMed]
32. Zsila, F.; Iwao, Y. The Drug Binding Site of Human Alpha1-acid Glycoprotein: Insight from Induced Circular Dichroism and Electronic Absorption Spectra. *Biochim. Biophys. Acta.* **2007**, *1770*, 797–809. [CrossRef] [PubMed]
33. Katsuki, M.; Chuang, V.T.; Nishi, K.; Kawahara, K.; Nakayama, H.; Yamaotsu, N.; Hirono, S.; Otagiri, M. Use of Photoaffinity Labeling and Site-directed Mutagenesis for Identification of the Key residue Responsible for Extraordinarily High Affinity Binding of UCN-01 in Human Alpha1-acid Glycoprotein. *J. Biol. Chem.* **2005**, *280*, 1384–1391. [CrossRef] [PubMed]
34. Becker, B.A.; Larive, C.K. Probing the Binding of Propranolol Enantiomers to α1-Acid Glycoprotein with Ligand-Detected NMR Experiments. *J. Phys. Chem. B* **2008**, *112*, 13581–13587. [CrossRef]
35. Miwa, T.; Ichikawa, M.; Tsuno, M.; Hattori, T.; Miyakawa, T.; Kayano, M.; Miyake, Y. Direct Liquid Chromatographic Resolution of Racemic Compounds. Use of Ovomucoid as a Column Ligand. *Chem. Pharm. Bull.* **1987**, *35*, 682–686. [CrossRef]
36. Miwa, T.; Miyakawa, T.; Kayano, M.; Miyake, Y. Application of an Ovomucoid-conjugated Column for the Optical Resolution of Some Pharmaceutically Important Compounds. *J. Chromatogr. A* **1987**, *408*, 316–322. [CrossRef]
37. Matsunaga, H.; Sadakane, Y.; Haginaka, J. Identification of Disulfide Bonds and Site-specific Glycosylation in Chicken Alpha1-acid Glycoprotein by Matrix-assisted Laser Desorption Ionization Time-of-flight Mass Spectrometry. *Anal. Biochem.* **2004**, *331*, 358–363. [CrossRef] [PubMed]
38. Matsunaga, H.; Haginaka, J. Investigation of Chiral Recognition Mechanism on Chicken Alpha1-acid Glycoprotein Using Separation System. *J. Chromatogr. A* **2006**, *1106*, 124–130. [CrossRef]
39. Zsila, F.; Matsunaga, H.; Bikádi, Z.; Haginaka, J. Multiple Ligand-binding Properties of the Lipocalin Member Chicken Alpha1-acid Glycoprotein Studied by Circular Dichroism and Electronic Absorption Spectroscopy: The Essential Role of the Conserved Tryptophan Residue. *Biochim. Biophys. Acta* **2006**, *1760*, 1248–1273. [CrossRef] [PubMed]
40. Chattopadhyay, A.; Tian, T.; Kortum, L.; Hage, D.S. Development of Tryptophan-modified Human Serum Albumin Columns for Site-specific Studies of Drug-protein Interactions by High-performance Affinity Chromatography. *J. Chromatogr. B* **1998**, *715*, 183–190. [CrossRef]
41. Haginaka, J.; Takehira, H. Separation of Enantiomers on a Chiral Stationary Phase Based on Ovoglycoprotein. I. Influences of the Pore Size of Base Silica Materials and Bound Protein Amounts on Chiral Resolution. *J. Chromatogr. A* **1977**, *773*, 85–91. [CrossRef]
42. Waterhouse, A.; Bertoni, M.; Bienert, S.; Studer, G.; Tauriello, G.; Gumienny, R.; Heer, F.T.; de Beer, T.; Rempfer, C.; Bordoli, L.; et al. SWISS-MODEL: Homology Modelling of Protein Structures and Complexes. *Nucleic Acids Res.* **2018**, *46*, W296–W303. [CrossRef] [PubMed]
43. Studer, G.; Tauriello, G.; Bienert, S.; Biasini, M.; Johner, N.; Schwede, T. ProMod3-A Versatile Homology Modelling Toolbox. *PLoS Comput. Biol.* **2021**, *17*, e1008667. [CrossRef]
44. Yang, J.; Zhang, Y. I-TASSER Server: New Development for Protein Structure and Function Predictions. *Nucleic Acids Res.* **2015**, *43*, W174–W181. [CrossRef] [PubMed]
45. Bharatham, N.; Bharatham, K.; Shelat, A.A.; Bashford, D. Ligand Binding Mode Prediction by Docking: Mdm2/Mdmx Inhibitors as a Case Study. *J. Chem. Inf. Model.* **2014**, *54*, 648–659. [CrossRef]
46. Kumar, G.K.; Prasanna, G.; Marimuthu, T.; Saraswathi, N.T. Structural Basis for Complementary and Alternative Medicine: Phytochemical Interaction with Non-structural Protein 2 Protease-a Reverse Engineering Strategy. *Chin. J. Integr. Med.* **2015**, *21*, 445–452. [CrossRef] [PubMed]
47. Ramakrishnan, P.; Pavan Kumar, T.; Saraswathy, G.R.; Sujatha, S. In Silico Evaluation of Drugs Used in Treatment of Oral Lichen Planus. *J. Oral Pathol. Med.* **2020**, *49*, 926–932. [CrossRef]
48. Friesner, R.A.; Banks, J.L.; Murphy, R.B.; Halgren, T.A.; Klicic, J.J.; Mainz, D.T.; Repasky, M.P.; Knoll, E.H.; Shelley, M.; Perry, J.K.; et al. Glide: A New Approach for Rapid, Accurate Docking and Scoring. 1. Method and Assessment of Docking Accuracy. *J. Med. Chem.* **2004**, *47*, 1739–1749. [CrossRef] [PubMed]
49. Haginaka, J.; Wakai, J.; Takahashi, K.; Yasuda, H.; Katagi, T. Chiral Separation of Propranolol and Its Ester Derivatives on an Ovomucoid-bonded Silica: Influence of pH, Ionic Strength and Organic Modifier on Retention, Enantioselectivity and Enantiomeric Elution Order. *Chromatographia* **1990**, *29*, 587–592. [CrossRef]

separations MDPI

Article

μQuEChERS Combined with UHPLC-PDA as a State-of-the-Art Analytical Approach for Quantification of Chlorpropham in Potato

Telma M. Gomes [1], Rosa Perestrelo [1] and José S. Câmara [1,2,*]

[1] CQM—Centro de Química da Madeira, PNRG, Universidade da Madeira, Campus da Penteada, 9020-105 Funchal, Portugal; telmamarisagomes@gmail.com (T.M.G.); rmp@staff.uma.pt (R.P.)
[2] Departamento de Química, Faculdade de Ciências Exatas e da Engenharia, Universidade da Madeira, Campus da Penteada, 9020-105 Funchal, Portugal
* Correspondence: jsc@staff.uma.pt; Tel.: +351-291-705-112

Abstract: Pesticides are chemicals used in agriculture to prevent insects, fungi, weeds, and other pests, from damaging crops. In addition, some types of pesticides are used after harvest as sprout suppressant agents help keeping the quality parameters of crops during storage. Nonetheless, its presence, even at trace levels, in food products is becoming a big challenge regarding human health. The current work aimed to develop and validate a sensitive and high-throughput analytical approach, based on a state-of-the-art microextraction technique—μQuEChERS, combined with ultra-high performance liquid chromatography equipped with a photodiode array detection system (UHPLC-PDA) to quantify isopropyl-N-(3-chlorophenyl) carbamate (chlorpropham), commonly used as efficient sprout suppressant stored potatoes, in raw and cooked potatoes cultivated in different geographical regions of Madeira Island (Portugal). Good results were obtained in terms of figures of the merit of the method, with correlation coefficients (R^2) higher than 0.999 and recoveries between 94.5% to 125%. Method limit of detection (LOD) and limit of quantification (LOQ) were 0.14 μg/Kg and 0.43 μg/Kg, respectively, which are much lower than the accepted and legislated requirements by the European Union, which is 20 μg/Kg for chlorpropham. The concentration of chlorpropham in raw potatoes is significantly higher when compared to cooked samples, which revealed that the thermic treatment during cooking had a significant effect on its degradation. A significant decrease (90%, on average) was observed on chlorpropham levels.

Keywords: herbicide; chlorpropham; potato; μQuEChERS/UHPLC-PDA; validation; cooking

Citation: Gomes, T.M.; Perestrelo, R.; Câmara, J.S. μQuEChERS Combined with UHPLC-PDA as a State-of-the-Art Analytical Approach for Quantification of Chlorpropham in Potato. *Separations* **2022**, *9*, 77. https://doi.org/10.3390/separations9030077

Academic Editor: Victoria Samanidou

Received: 5 March 2022
Accepted: 12 March 2022
Published: 17 March 2022

1. Introduction

The potato (*Solanum tuberosum* L.) is a tuber produced by the potato plant, an herbaceous perennial plant belonging to the *Solanaceae* family. It consists of about 80% water, 7% carbohydrates (88% is starch), 2% protein, and negligible amounts of fat, depending on the cultivar. It is a rich source of vitamin C and a moderate source of vitamins from the B complex, minerals, fibers, and antioxidants, which prevents aging-related diseases [1,2]. The world production is led by China (20–22%), followed by India, Russia, Ukraine, and the United States [1,2]. The species originated in the Andes Mountains, near Lake Titicaca, and was taken to other regions of the world by European settlers. It is the fourth most consumed crop in the world, surpassed only by rice, wheat and maize, and remains an essential crop in Europe, especially in eastern and northern regions [2].

Like several other crops, potato plantations are subject to the attack of various species of bacteria, fungi, and insects that compromise their productivity. Due to its importance expressed into its high consumption worldwide (the global importance of potatoes is so great that FAO, the UN body for Agriculture and Food instituted 2008, as the "*International Year of Potatoes*"), it is essential to create more resistant varieties, contributing to an increase in productivity and thus minimizing hunger in several countries. The scarcity of these resistant varieties requires the use of pesticides to control pests and contribute to increase productivity [1,3,4].

Depending on the degree of toxicity, each pesticide has a maximum permissible value of the application, which should not be exceeded to minimize the harmful effects on human health. Exposure to high amounts of pesticides can cause neurological and oncological disorders, among others [5,6]. Generally, the maximum residue limits for most pesticides are variable depending on their toxicity. For glyphosate and its metabolite [7], the maximum value is 0.01 mg/kg.

Herbicides can be classified as (i) selective, which inhibit or slow the growth of weeds by keeping the desired vegetation intact, and (ii) non-selective ones, which destroy any form of plant life. The selective ones can differentiate the vegetation of interest from the unwanted due to its mode of action, inactivating the enzymatic action of unwanted plants and the metabolic processes associated with their development [8,9]. On the other hand, non-selective herbicides act more generally and can also act through enzymes, modifying those that are common in all plant species. They can also act by photosynthesis through the solar energy captured by chlorophyll. This deviation of the flow of electrifications through photosystem I result in the production of free radicals, which, as they are very reactive, end up destroying cell membranes, leaving the leaves of plants yellow and dry. Another mode of action is processed by blocking the energy associated with photosystem II, where the herbicide binds to the plastoquinone protein (involved in the carrier chain of electrodes), reducing its effectiveness, thus slowing the growth of the plant due to the reduction of available energy from photosynthesis. Non-selective herbicides can also modify the cell cycle by inhibiting the development of meristematic cells, preventing cell division [9].

Despite the efficiency of herbicides in the control of pests and the consequent increase in crop productivity, they have been the target of attention from the scientific and medical communities due to their potentially harmful effects on human health. BATALEX, which has as active substance the isopropyl-N-(3-chlorophenyl) carbamate (chlorpropham), is amongst the herbicides commonly used as anti-sprouting agent, to which are associated toxicological and carcinogenic effects. Chlorpropham is an organic pesticide belonging to the carbamate group (functional group —NHCOO), being chemically classified as a carbamic acid (NH_2COOH). The generic reaction of carbamate synthesis is represented in Figure 1 [10].

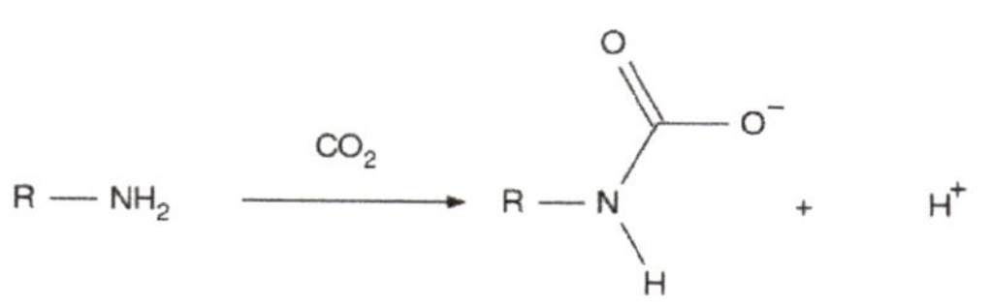

Figure 1. Generic reaction of carbamate synthesis.

The synthesis of carbamates may occur: (a) by reaction of an amine, with a carbonyl group and alcohol; (b) from the reaction between amines with chloroformate (alcohols or phenols reaction products with phosgene); and (c) from the reaction of alcohol or phenols with isocyanates obtained from the phosgene reaction with amines, shown in Figure 2 [11].

Chlorpropham inhibits the formation of meristematic cells, being widely used as anti-sprouting agent (inhibits the formation of turnip greens) in potatoes, thus contributing to increasing its shelf life with high quality parameters. In addition, this action inhibits the release of α-solanine and α-chaconine, which are harmful for health.. Meristematic cells are totipotent cells located in regions where plant growth occurs; they are undifferentiated cells that have great multiplication capacity and to differentiate in any cell type [12,13]. Chlorpropham is classified as a medium toxicity pesticide (toxicological class II). Currently, its use is authorized in the European Union countries and is legislated as the maximum limit of 20 μg/kg. The lethal dose (LD_{50}) of chlorpropham is 3.80 mg/kg [6].

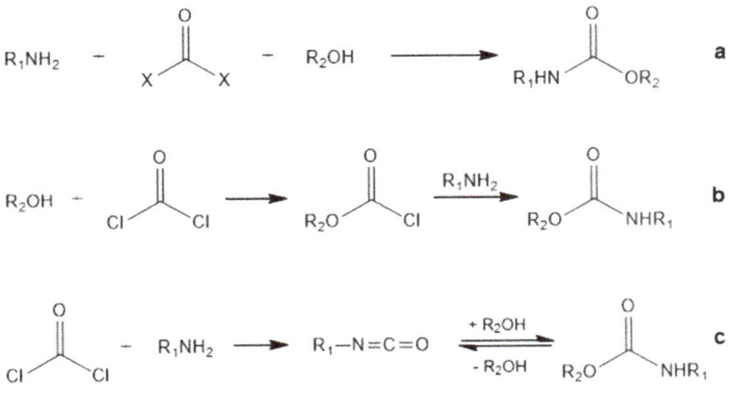

Figure 2. Carbamate synthesis. (**a**) reaction between amines, carbonyl group and alcohol; (**b**) reaction chloroformat with amines; (**c**) isocyanate reaction with alcoholics or phenols.

The low concentrations of pesticides residues in foods make difficult its direct quantification by chromatographic methods, such as gas chromatography (GC) or liquid chromatography (LC) [14]. Consequently, it is necessary to apply an sample preparation techniques for the extraction and preconcentration of the pesticides or other residues from foods or other samples. In this sense, different extraction procedures have been recommended to extract herbicides from foods, such as vortex-assisted liquid–liquid microextraction (VALLME) [15], solid-phase extraction (SPE) [16], solid-phase microextraction (SPME) [17], dispersive liquid–liquid microextraction (DLLME) [14], among others. These extraction procedures are expensive, labor-intensive, and time-consuming. Today, a quick, easy, cheap, effective, rugged and safe method (QuEChERS), followed by clean-up steps involving dispersive solid-phase extraction (dSPE), is one of the most promising user-friendly and high throughput extraction procedures, using low solvent and sample amounts to extract pesticides from complex matrices, providing high-quality results with a reduced number of steps [18,19].

The current work aimed to validate and apply a state-of-the-art, quick, easy, cheap, effective, rugged, and safe microextraction technique followed by dispersive solid-phase extraction-based clean-up (μQuEChERS-dSPE) combined with ultra-high performance liquid chromatography equipped with a photodiode array detection system (UHPLC-PDA) for the quantification the anti-sprout agent, chlorpropham, in different parts of potato (potato skin, pulp, and whole potato). The influence of cooking on chlorpropham levels was also evaluated.

2. Materials and Methods

2.1. Chemicals

All solvents and other chemicals were of analytical quality grade. HPLC grade acetonitrile (MeCN) was obtained from LabScan (Dublin, Ireland). Herbicide standard, chlorpropham (98%), as well as the buffered salts used in QuEChERS extraction–sodium chloride (NaCl), anhydrous magnesium sulfate ($MgSO_4$), disodium hydrogen citrate sesquihydrate ($C_6H_8Na_2O_8$), and trisodium citrate dihydrate ($C_6H_5Na_3O_7 \cdot 2H_2O$), were supplied by Sigma-Aldrich (St. Louis, MO, USA). dSPE clean-up DisQuETM tubes with primary, secondary amine (PSA), $MgSO_4$, and C_{18} were obtained from Waters (Milford, MA, USA). Formic acid (FA, ≥99%) and acetic acid (≥99%) were supplied from Merck (Darmstadt, Germany). Ultrapure water (H_2O) from a Milli-Q ultrapure water purification system (Millipore, Bedford, MA, USA) was used for preparing the UHPLC mobile phase. Before UHPLC-PDA analysis, the final extracts were filtered through 13 mm with 0.22-μm PTFE membranes.

2.2. Sample Preparation

The potato samples (1 Kg) were provided by the Agricultural Markets of Porto Moniz (Asterix variety), Prazeres, and Santana (Desire variety). From each sample, the following potato constituents: skin, pulp, and potato (skin + pulp) were analyzed raw and cooked.

2.3. Standard Solution

Individual stock solution of chlorpropham standard was prepared at a concentration of 400 mg/L in MeCN containing 0.1% of acetic acid and stored at −20 °C in the dark for a maximum of six months. Intermediate stock solution at 5 mg/L of chlorpropham was prepared in MeCN. The working standard solutions used to construct the calibration curve were prepared by the appropriate dilution of aliquots of the intermediate stock solution in MeCN to obtain the concentration range of 1–200 μg/Kg. The density was used to convert mg/L to μg/Kg. All standard solutions were labeled and stored at −20 °C.

2.4. μQuEChERS-dSPE Procedure

For μQuEChERS extraction, 500 μg of sample was weighed to the accuracy of 0.0001 mg, put into a centrifuge tube of 5 mL polytetrafluoroethylene (PTFE), mixed, and left to stand for 15 min at room temperature. Afterward, 1000 μL MeCN was added, and the resulting mixture was shaken vigorously for 1 min with a vortex ensuring that the solvent interrelated well with the entire sample. Buffer salts, $MgSO_4$, NaCl, $C_6H_5Na_3O_7{\cdot}2H_2O$ and $C_6H_8Na_2O_8$, in proportions of 4:1:1:0.5, were added to the homogenized mixture, vortexed for 2 min and centrifuged at 4000 rpm for 3 min at 25 ± 1 °C, ending the partition step and the consequent separation of phases (aqueous and organic phase) [19]. For removal of potentially interfering compounds, 500 μL of the supernatant was placed in a DisQuETM dSPE clean-up tube containing 2.5 mg of PSA, 15 mg of $MgSO_4$, and 2.5 mg of C18, to remove proteins, lipids, and other interferences. The mixture was centrifuged (4000 rpm, 3 min, 25 °C), and 200 μL of supernatant was filtered through a 0.22-μm PTFE filter membrane to a vial for UHPLC-PDA analysis.

2.5. UHPLC-PDA Conditions

The separation, identification, and quantification of chlorpropham was performed on a Waters Ultra Pressure Liquid Chromatographic Acquity system (UPLC, Acquity H-Class) (Milford, MA, USA) coupled with a Waters Acquity quaternary solvent manager (QSM), an Acquity sample manager (SM), a column heater, a 2996 PDA detector and a degassing system. The whole configuration was controlled by Empower software v2.0 from Waters Corporation. The used column for chlorpropham separation was CORTECS UPLC C18 (2.1 × 100 mm, 1.6 μm) maintained at a temperature of 30 °C. The mobile phase was composed of H_2O with 0.1% formic acid (solvent A) and MeCN (solvent B) according to the following gradient: isocratic at 10% B from 0 to 3 min, from 3 to 4 min gradient from 10 to 50% B, gradient from 50 to 65% B from 4 to 10 min, gradient from 65 to 70% from 10 to 11 min, and finally from 70 to 75% from 11 to 19 min. Following the system, a return to the initial mobile phase composition from 75% to 10% was within 1 min. Five μL of extract was injected at constant flow of 350 μL/min.

For quantification purposes, the PDA detection was performed at 235 nm. The chlorpropham was identified by comparing its retention time (RT) and spectral features obtained for an extract with those of the pure standard. The quantification was carried out by means of the chlorpropham standard in triplicate. The results were presented as mean ± standard deviation. The peak purity was confirmed peak by the screening of the chlorpropham UV spectrum from the beginning to the end of the peak.

2.6. Method Validation

The μQuEChERS method was validated concerning the linearity, limit of detection (LOD), limit of quantification (LOQ), precision (intra-day and inter-day), matrix effect, and recovery, according to European Union SANCO/12495/2011 guidelines.

The absence of interfering peaks at the chlorpropham RT, is used to assess the method selectivity and correspond to the level to which a method can quantify a specific analyte in a complex mixture without interference from other analytes.

Nine-points calibration curve was constructed with the following chlorpropham concentrations: 200, 150, 100, 50, 25, 10, 5.0, 2.5, and 1.0 μg/Kg, to determine the method's linearity. As part of the method linearity assessment, linearity range and determination coefficients (R^2) were evaluated.

The LOD, the lowest concentration of analyte that can be detected, and the LOQ, the lowest quantity of analyte that can be calculated quantitatively with satisfactory precision and accuracy, under the stated operating conditions of the method, were used to evaluate the method sensitivity. These parameters were determined by using the residual standard deviation ($S_{y/x}$) of corresponding curves being LOD and LOQ calculated by 3.3 $S_{y/x}/b$ and 10 $S_{y/x}/b$, respectively, where b represents the slope of calibration curve.

Recovery was evaluated at three concentration levels (in triplicate) within the linear range of the calibration curve, which allowed to evaluate the method accuracy. It is calculated through the relation of theoretical concentration added to the sample ($C_{theoretical}$) to the experimental concentration ($C_{experimental}$) of chlorpropham in the sample. The $C_{experimental}$ was calculated by the variation between the peak area of the chlorpropham in spiked and non-spiked samples.

Intra- and inter-day precision, expressed as percentage of relative standard deviation (% RSD), was calculated from triplicate assays of sample spiked at three concentration levels during the same day (repeatability) and in five consecutive days (reproducibility).

The matrix effect, most noticeable in complex samples, was determined based on the method of "standard additions" applied to the sample investigated, which was measured by the correlation of the slopes from the calibration curve of the chlorpropham in sample and in the solvent-based matrix.

2.7. Statistical Analysis

The multivariate data analysis (MVDA) was performed using the MetaboAnalyst 5.0 web-based tool (Chong et al., 2018; Pang et al., 2021). The data obtained were normalized (data transformation by cubic root and data scaling by auto-scaling). Then, principal component analysis (PCA) (Figure S1, supplementary material) and partial least squares-discriminant analysis (PLS-DA) were used to provide insights into separations among the samples under study.

3. Results and Discussion

3.1. Method Validation

The performance of μQuEChERS/UHPLC-PDA was evaluated for selectivity, linearity, accuracy (% recovery), precision (intra- and inter-day), and sensitivity (LOD and LOQ).

The selectivity was determined through the nonappearance of interfering peaks at the RT and wavelength of the studied analyte by the evaluation of the matrix effect. In order to evaluate the matrix effect, μQuEChERS extractions were performed in fortified potatoes with the same concentrations used previously. The nonappearance of chlorpropham in the sample was previously confirmed. The matrix effect was calculated by the slope ratio of the calibration line obtained with the sample by the slope of the calibration line obtained in the solvent. No significant interference was observed at the RT of chlorpropham in the potato matrix, which confirms the method selectivity.

The method linearity was evaluated through calibration curves that fit the least square linear regression analysis model. The correlation coefficient (R^2) obtained was 0.999, with residuals lower than ±15%, which indicates an excellent linear relationship between area vs. chlorpropham concentration. Additionally, the μQuEChERS-dSPE/UHPLC-PDA analytical methodology shows a big potential to detect and quantify the chlorpropham since the LOD was 0.14 μg/Kg and the LOQ was 0.43 μg/Kg.

The precision and recovery were evaluated by spiking potato samples with chlorpropham at different concentration levels (Table 1) within the linear range of the calibration curve. The intra-day precision ranged from 1.5% to 14.6%, while the inter-day precision from 4.2% to 16.3%. The recovery of chlorpropham ranged from 94.5% to 125%. According to the literature, a quantitative method should be validated as being able to show a mean recovery from 70% to 120%, and its precision should show % RSD values lower than 20%. Nevertheless, for concentrations at 1.0 μg/Kg, the recovery is slightly higher (125%); however, this value is within the analytical error allowed (120 ± 5%).

Table 1. Recovery and precision for chlorpropham quantification using μQuEChERS-dSPE/UHPLC-PDA methodology.

Concentration Range (μg/Kg)		Precision (%RSD)		Accuracy
Theoretical	Experimental	Intra-Day	Inter-Day	Rec (%) ± SD
200	211	1.5	4.2	106 ± 3.06
150	157	3.4	5.1	105 ± 4.26
100	103	4.5	6.8	103 ± 4.09
50.0	49.1	8.5	9.4	95.7 ± 2.98
25.0	24.2	6.8	10.2	94.5 ± 1.43
10	9.6	7.6	9.7	95.9 ± 0.98
5	5.4	10.4	11.6	112 ± 3.47
2.5	2.7	12.2	13.5	118 ± 7.09
1	1.6	14.6	15.7	125 ± 8.03

Moreover, the developed analytical method was compared with other liquid chromatography (LC) and gas chromatography (GC) methods reported in the literature for the quantification of chlorpropham in vegetables and water samples (Table 2) [15,16,20,21]. The low sample amount (g), LODs, LOQs, and recovery were assessed to prove the benefits of the μQuEChERS-dSPE/UHPLC-PDA method. The current analytical method proposed with this study used the lowest sample amount (500 μg).in comparison with other methods reported in studies to quantify the chlorpropham. Nonetheless, the VALLME/HPLC-AD and SPE methods require large solvent volumes compared to μQuEChERS-dSPE. μQuEChERS-dSPE/UHPLC-PDA showed enhanced analytical performance compared to most of the reference methods.

Table 2. Comparison of the analytical parameters of few studies, reported in the literature, for quantification of chlorpropham in different samples.

Sample (Amount)	Extraction Procedure	Analytical Method	LOD (μg/Kg)	LOQ (μg/Kg)	Rec (%)	Ref.
Water (18 mL)	SPME	GC-MS/MS	0.02	0.06 *	95.3–98.9	[20]
Potatoes (0.5 g)	VALLME	HPLC-AD	3.67	12.2 *	75.7–104	[15]
Potatoes (5 g)	SPE	HPL-UV	30	100	90.7–97.0	[16]
Carrot, tomatoes (5 g)	QuEChERS-dSPE	HPLC-UV	1077	3589	51.3–73.0	[21]
Potatoes (0.5 mg)	***μQuEChERS-dSPE***	***UPLC-PDA***	***0.14***	***0.43***	***94.5–125***	***This work***

Abbreviations—GC-MS/MS: gas chromatography tandem mass spectrometry; HPLC-AD: high-performance liquid chromatography with amperometry detection; HPLC-UV: high-performance liquid chromatography with ultraviolet detector; LOD: limit of detection; LOQ: limit of quantification; QuEChERS-dSPE: quick, easy, cheap, effective, rugged and safe technique coupled with dispersive solid-phase extraction; Rec (%): recovery; SPE: solid-phase extraction; SPME: solid-phase microextraction; UPLC-PDA: ultrahigh performance liquid chromatography coupled with photodiode array detector; VALLME: vortex-assisted liquid–liquid microextraction. * Expressed as μg/L.

3.2. Quantification of Chlorpropham in Potatoes

Chlorpropham was determined in different constituent parts of raw and cooked potato—outside (skin), inside (pulp), whole potato (potato skin + pulp), and in cooking water. The potato samples were provided by the Agriculture Market from Porto Moniz,

Prazeres, and Santana. The chromatograms of the chlorpropham standard and the samples are presented in Figure 3.

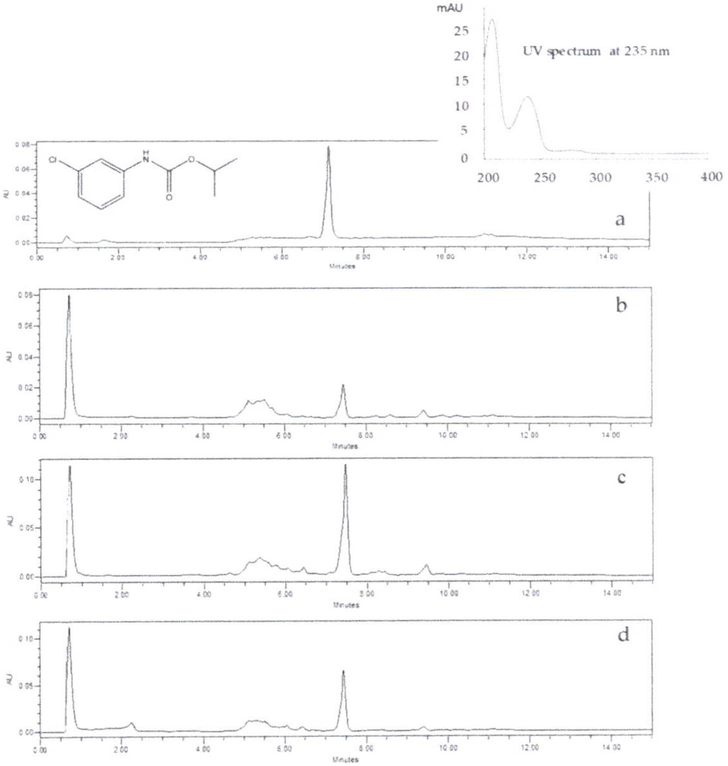

Figure 3. UPLC UV/VIS chromatograms of a chlorpropham standard solution (**a**) and chlorpropham in the potato skin samples from the agricultural markets of Porto Moniz (**b**), Prazeres (**c**) and Santana (**d**) obtained by μQuEChERS/UHPLC-PDA at λ = 235 nm.

The concentration of chlorpropham in the raw potato skins from Porto Moniz, Prazeres, and Santana, was 15.4 μg/Kg, 85.4 μg/Kg, and 40.7 μg/Kg, respectively. The levels found in the raw potato skins from Prazeres and Santana are above the ceiling accepted and legislated by the EU, 20 μg/Kg. Nevertheless, after cooking, the concentration of chlorpropham in these samples was significantly reduced to values lower than 1 μg/Kg, which indicates that the use of chlorpropham does not represent a danger to public health. On the other hand, in the whole potato (potato skin + pulp) and in the pulp samples, the values presented are much lower than those regulated (Table 3). The presence of chlorpropham in the potato pulp indicates that there was a transfer of mass by diffusion from potato skin to the pulp. The presence of chlorpropham in the potatoes cooking water was also analyzed, and the herbicide was identified in concentrations ranging from 1.04 μg/kg (cooking water of Prazeres potatoes) to 0.71 μg/Kg (Santana).

As can be seen from the results presented in Table 3, the concentration of chlorpropham in raw potato skin and potato samples (potato skin + pulp) is significantly higher when compared to cooked samples meaning that the thermic treatment during cooking had a significant effect on the decomposition of chlorpropham since its concentration was reduced by about 90%, on average. On the other hand, the concentration of chlorpropham in raw pulp (0.74 μg/Kg, on average) does not differ significantly ($p < 0.05$) from the values found in cooked pulp (0.96 μg/Kg, on average).

Table 3. Concentration (μg/Kg) ± standard deviation of chlorpropham in raw and cooked potatoes and in cooking water.

Samples	Porto Moniz	Prazeres	Santana
		Raw	
Potato skin	15.4 ± 1.02	85.4 ± 2.43	40.7 ± 1.47
Potato pulp	0.64 ± 0.01	0.80 ± 0.01	0.78 ± 0.01
Potato (skin + pulp)	7.39 ± 0.04	15.4 ± 0.87	8.31 ± 0.03
		Cooked	
Potato skin	0.72 ± 0.02	0.79 ± 0.02	0.85 ± 0.03
Potato pulp	0.81 ± 0.03	1.18 ± 0.05	0.88 ± 0.05
Potato (skin + pulp)	1.21 ± 0.02	0.68 ± 0.01	0.69 ± 0.04
Cooking water	1.03 ± 0.01	1.04 ± 0.01	0.71 ± 0.01

To further understand the differences between raw and cooked potatoes, a PLS-DA model was developed. Nevertheless, the score plot (Figure 4a) did not show a clear separation of potatoes based on the raw and cooked process. Therefore, a novel OPLS-DA model was built, and clear discrimination was observed among potatoes cultivated in different geographical regions of Madeira Island–Porto Moniz, Prazeres, and Santana (Figure 4b). Three significant components described 0.7982 of the goodness of fit (R^2 = 79.82%) and predicted ability of 0.6921 (Q^2 = 69.21%) based on crossing-validation.

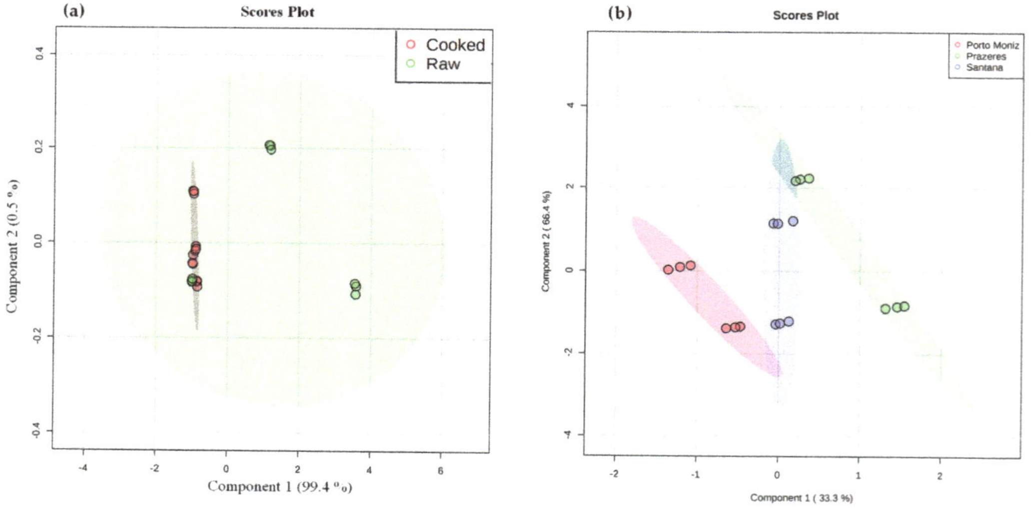

Figure 4. (**a**) Partial least square-discrimination analysis (PLS-DA) of raw and cooked of different parts of potatoes, and (**b**) Partial least square-discrimination analysis (PLS-DA) of different parts of potatoes by geographical regions.

4. Conclusions

A simple, fast, cheap, and high-throughput μQuEChERS-dSPE followed by UHPLC-PDA was developed and validated to quantify chlorpropham in different parts of potatoes. Satisfactory figures of merit of the method were attained in terms of linearity ($R^2 \geq 0.999$), intra-day/inter-day precision (RSD < 16%), recovery (94.5–125%), and sensitivity (low

LOD and LOQ) for chlorpropham. In the potato skin before cooking, extremely high chlorpropham values were obtained, especially in the sample from Santana (40.7 μg/Kg) and Prazeres (85.4 μg/Kg), which exceeded the accepted and legislated by the EU limit, 20 μg/Kg. For the remaining parts of potatoes, independently if raw and cooked, the concentration of chlorpropham was lower than legislated by the EU. Regarding the influence of cooking on chlorpropham concentration on the different constituent parts of the potato, there was a significant decrease (90% on average) in the chlorpropham concentration in the potato skin of the cooked potato.

The μQuEChERS-dSPE revealed a suitable green and state-of-the-art microextraction technique for routine practice since it is simple, cheap, accurate, precise, and environmentally friendly. In addition to UHPLC-PDA analysis, it constitutes a high throughput separation technique with a high-resolution power in a short run time, which makes the μQuEChERS-dSPE/UHPLC-PDA methodology a useful approach for its application to other types of pesticides and food matrices. After cooking, the concentration of chlorpropham in potatoes was lower than the maximum residue limits (MRLs) set by the EU, which means that the cooking process has a significant impact on the degradation of chlorpropham. The obtained results revealed that the use of chlorpropham in potatoes as an anti-sprouting agent does not constitute any risk to human health, provided that the concentration levels determined for the analyzed samples are below their MRL value.

Supplementary Materials: The following supporting information can be downloaded at: https://www.mdpi.com/article/10.3390/separations9030077/s1, Figure S1. (a) Principal component analysis (PCA) of raw and cooked of different parts of potatoes, and (b) Principal component analysis (PCA) of different parts of potatoes by geographical regions.

Author Contributions: Conceptualization, J.S.C.; methodology, T.M.G.; formal and analysis, T.M.G.; investigation, T.M.G. and R.P.; resources, J.S.C.; data curation, T.M.G., R.P. and J.S.C.; writing—original draft preparation, T.M.G. and R.P.; writing—review and editing, J.S.C.; supervision, J.S.C. All authors have read and agreed to the published version of the manuscript.

Funding: This research was funded by FCT-Fundação para a Ciência e a Tecnologia through the CQM Base Fund, UIDB/00674/2020, Programmatic Fund, UIDP/00674/2020, and by ARDITI-Agência Regional para o Desenvolvimento da Investigação Tecnologia e Inovação, through the project M1420-01-0145-FEDER-000005—Centro de Química da Madeira—CQM+ (Madeira 14-20 Program). The authors also acknowledge FCT and Madeira 14-2020 program to the Portuguese Mass Spectrometry Network (RNEM) through PROEQUIPRAM program, M14-20 M1420-01-0145-FEDER-000008).

Institutional Review Board Statement: Not applicable.

Informed Consent Statement: Nor applicable.

Acknowledgments: The authors acknowledge Victoria F. Samanidou, from Aristotle University of Thessaloniki for the kind invitation to participate in this Special Issue.

Conflicts of Interest: The authors declare no conflict of interest.

References

1. Chen, L.; Jia, C.; Li, F.; Jing, J.; Yu, P.; He, M.; Zhao, E. Dissipation and residues of fluazinam and dimethomorph in potatoes, potato plants, and soil, determined by QuEChERS ultra-performance liquid chromatography tandem mass spectrometry. *Environ. Sci. Pollut. Res.* **2018**, *25*, 32783–32790. [CrossRef] [PubMed]
2. Reis, D.; Silva, P.; Perestrelo, R.; Cmara, J.S. Residue analysis of insecticides in potatoes by QuEChERS-dSPE/UHPLC-PDA. *Foods* **2020**, *9*, 1000. [CrossRef]
3. Agatz, A.; Ashauer, R.; Sweeney, P.; Brown, C.D. A knowledge-based approach to designing control strategies for agricultural pests. *Agric. Syst.* **2020**, *183*, 102865. [CrossRef] [PubMed]
4. Nicolopoulou-Stamati, P.; Maipas, S.; Kotampasi, C.; Stamatis, P.; Hens, L. Chemical Pesticides and Human Health: The Urgent Need for a New Concept in Agriculture. *Front. Public Health* **2016**, *4*, 148. [CrossRef]
5. Casida, J.E.; Bryant, R.J. The ABCs of pesticide toxicology: Amounts, biology, and chemistry. *Toxicol. Res.* **2017**, *6*, 755–763. [CrossRef] [PubMed]
6. Richardson, J.R.; Fitsanakis, V.; Westerink, R.H.S.; Kanthasamy, A.G. Neurotoxicity of pesticides. *Acta Neuropathol.* **2019**, *138*, 343–362. [CrossRef] [PubMed]

7. Silva, V.; Mol, H.G.J.; Zomer, P.; Tienstra, M.; Ritsema, C.J.; Geissen, V. Pesticide residues in European agricultural soils—A hidden reality unfolded. *Sci. Total Environ.* **2019**, *653*, 1532–1545. [CrossRef] [PubMed]
8. Long, C.S.; Novelo, N.N.; Derecho, H.; Youngblood, Y. The Effects of a Selective and Non-Selective Organic Herbicides on Amaranthus species. *Pursue* **2018**, *1*, 50–62.
9. Sherwani, S.I.; Arif, I.A.; Khan, H.A. Modes of Action of Different Classes of Herbicides. In *Herbicides: Physiology of Action and Safety*; InTech: Rijeka, Croatia, 2015. [CrossRef]
10. Gupta, P.K. *Toxicity of Herbicides*, 3rd ed.; Elsevier Inc.: Amsterdam, The Netherlands, 2018; ISBN 9780128114100.
11. Ghosh, A.K.; Brindisi, M. Organic Carbamates in Drug Design and Medicinal Chemistry. *J. Med. Chem.* **2015**, *58*, 2895–2940. [CrossRef]
12. Arena, M.; Auteri, D.; Barmaz, S.; Bellisai, G.; Brancato, A.; Brocca, D.; Bura, L.; Byers, H.; Chiusolo, A.; Court Marques, D.; et al. Peer review of the pesticide risk assessment of the active substance chlorpropham. *EFSA J.* **2017**, *15*, e04903. [CrossRef]
13. Göckener, B.; Kotthoff, M.; Kling, H.W.; Bücking, M. Fate of Chlorpropham during High-Temperature Processing of Potatoes. *J. Agric. Food Chem.* **2020**, *68*, 2578–2587. [CrossRef] [PubMed]
14. Wang, Y.; Sun, Y.; Xu, B.; Li, X.; Jin, R.; Zhang, H.; Song, D. Magnetic ionic liquid-based dispersive liquid–liquid microextraction for the determination of triazine herbicides in vegetable oils by liquid chromatography. *J. Chromatogr. A* **2014**, *1373*, 9–16. [CrossRef]
15. Diuzheva, A.; Dejmková, H.; Fischer, J.; Andruch, V. Simultaneous determination of three carbamate pesticides using vortex-assisted liquid–liquid microextraction combined with HPLC-amperometric detection. *Microchem. J.* **2019**, *150*, 104071. [CrossRef]
16. Douglas, L.; MacKinnon, G.; Cook, G.; Duncan, H.; Briddon, A.; Seamark, S. Determination of chlorpropham (CIPC) residues, in the concrete flooring of potato stores, using quantitative (HPLC UV/VIS) and qualitative (GCMS) methods. *Chemosphere* **2018**, *195*, 119–124. [CrossRef] [PubMed]
17. Berrada, H.; Font, G.; Moltó, J.C. Application of solid-phase microextraction for determining phenylurea herbicides and their homologous anilines from vegetables. *J. Chromatogr. A* **2004**, *1042*, 9–14. [CrossRef]
18. Perestrelo, R.; Silva, P.; Porto-Figueira, P.; Pereira, J.A.M.J.A.M.; Silva, C.; Medina, S.; Câmara, J.S. QuEChERS—Fundamentals, relevant improvements, applications and future trends. *Anal. Chim. Acta* **2019**, *1070*, 1–28. [CrossRef] [PubMed]
19. Sousa, J.F.; Barros, J.N.; Fernandes, P.; Perestrelo, R.; Câmara, J.S. Simultaneous determination of N-methyl carbamate residues in pork tissues based on ultrasound assisted QuEChERS-dSPE extraction followed by reversed phase LC-FLD analysis. *LWT* **2021**, *144*, 111199. [CrossRef]
20. Jabali, Y.; Millet, M.; El-Hoz, M. Optimization of a DI-SPME-GC–MS/MS method for multi-residue analysis of pesticides in waters. *Microchem. J.* **2019**, *147*, 83–92. [CrossRef]
21. Sayed, A.; Chys, M.; De Rop, J.; Goeteyn, L.; Spanoghe, P.; Sampers, I. Pesticide residues in (treated) wastewater and products of Belgian vegetable- and potato processing companies. *Chemosphere* **2021**, *280*, 130619. [CrossRef] [PubMed]

separations

MDPI

Article

An Integrated Chromatographic Strategy for the Large-Scale Extraction of Ergosterol from *Tulasnellaceae* sp.

Ze Wang [1], Qi Wang [2], Jinkui Zhang [2], Pengcheng Lin [1,*] and Jun Dang [2,*]

1 Key Laboratory of Tibet Plateau Phytochemistry of Qinghai Province, Qinghai Nationalities University, Xining 810007, China; qhzewang@126.com

2 Key Laboratory of Tibetan Medicine Research, Northwest Institute of Plateau Biology, Chinese Academy of Sciences, Xining 810007, China; wqqwe1996@163.com (Q.W.); qhmuzjk@126.com (J.Z.)

* Correspondence: qhlpc@126.com (P.L.); dangjun@nwipb.cas.cn (J.D.); Tel.: +86-971-880-4649 (P.L.); +86-971-614-3282 (J.D.)

Abstract: A reliable chromatographic strategy is crucial for the extraction of target compounds from natural sources as it is related to the preparation efficiency, as well as the purity of the compounds. In this study, medium-pressure normal-phase liquid chromatography and high-pressure reverse-phase liquid chromatography were combined to prepare and purify ergosterol from *Tulasnellaceae* sp. of *Gymnadenia orchidis*. First, *Tulasnellaceae* sp. was extracted three times (2.0 L and 2 h each time) with ethyl acetate, and the 6.0 L of extract solution was concentrated under reduced pressure to yield 2.2 g of crude sample. Then, the crude sample was pretreated utilizing silica gel medium-pressure liquid chromatography to enrich the target ingredient (586.0 mg). Finally, high-pressure reversed-phase liquid chromatography was used to purify the target compound, and the compound was characterized as ergosterol (purity > 95%) using spectral data. Overall, the simple and reproducible integrated chromatographic strategy developed in this study has the potential for the large-scale purification of steroids for laboratory and even industrial research. To the best of our knowledge, this is also the first report of ergosterol in *Tulasnellaceae* sp.

Keywords: *Tulasnellaceae* sp.; *Gymnadenia orchidis*; mdium-pressure liquid chromatography; reversed-phase liquid chromatography; ergosterol

Citation: Wang, Z.; Wang, Q.; Zhang, J.; Lin, P.; Dang, J. An Integrated Chromatographic Strategy for the Large-Scale Extraction of Ergosterol from *Tulasnellaceae* sp. *Separations* **2022**, *9*, 176. https://doi.org/10.3390/separations9070176

Academic Editor: Victoria Samanidou

Received: 29 June 2022
Accepted: 14 July 2022
Published: 17 July 2022

1. Introduction

Natural products continue to attract scientists' attention due to their abundant sources and broad range of biological functions [1]. *Gymnosia orchidis* (also known as "Wanglag") is a kind of Tibetan medicine, which is used to treat kidney deficiency, physical weakness, and pain [2]. It is widely found in Yunnan, Tibet, Sichuan, Qinghai, Shanxi, and other places in China [3–5]. The Chinese government has currently placed it on the national second-class protected plant list. *Tulasnellaceae* sp. is one of the most significant endophytic fungi in *Gymnadenia orchidis*. Recent studies have demonstrated that ergosterol is the major component of *Gymnadenia orchidis*'s endogenous metabolite. It is a characteristic sterol of fungi, and has the functions of protecting the liver, antifibrosis, and inhibiting tumor cells. However, establishing its quality standard and evaluating its biological activity have been challenging due to a lack of adequate chromatographic purification techniques [6–8]. Therefore, the isolation of ergosterol can lay the foundation for subsequent quality control research and pharmacological activity research.

For the separation and purification of steroids (such as ergosterol) from plants, conventional methods such as thin-layer chromatography and silica gel column chromatography are commonly utilized [9–11] as these procedures only require basic equipment and are straightforward to implement. The separation of high-purity chemicals, however, continues to be problematic due to low separation repeatability and the lack of online detection methods. High-speed countercurrent chromatography (HSCCC) has recently gained popularity for the separation of ergosterol from microbial samples and natural products [10,11], but HSCCC requires complex partition coefficient testing procedures and low separation resolution, making the separation and purification process difficult to simplify [12–14].

Preparative high-performance liquid chromatography (prep-HPLC) is an efficient strategy for the large-scale separation of individual constituents from complex samples, including microorganisms and natural products [15–18]. This method is extensively employed in numerous areas because of its excellent column efficiency, superior separation reproducibility, online detection, and automatic control [19,20]. Usually, it is worthwhile to pretreat the sample by medium-pressure liquid chromatography prior to HPLC preparation to reduce column contamination and enrich for target ingredients. Silica gel, resin, and polyamide are commonly used as sample pretreatment fillers due to their low cost and high separation selectivity [21,22]. Currently, several researchers have shown the viability of silica gel in the separation of steroids [23,24].

In view of the many reports mentioned above, in this study, we combined silica gel medium-pressure liquid chromatography with high-pressure liquid chromatography to establish a normal-phase medium-pressure/reverse-phase high-pressure mode for the efficient separation and purification of ergosterol from *Tulasnellaceae* sp. The final experimental findings revealed that such an integration strategy is economical and repeatable, and meets experimental predictions. It offers a methodological guide for subsequent attempts to produce ergosterol and its structural analogs from natural sources. Simultaneously, i this is also the first description of ergosterol in *Tulasnellaceae* sp.

2. Materials and Methods

2.1. Apparatus and Reagents

An industrial-level sample pretreatment chromatographic apparatus consisting of two NP7000 pre-HPLC pumps (Jiangsu Hanbon Science & Technology, Hanbon, China), a UV detector, a manual injector (5.0 mL maximum injection volume), and an EasyChrom workstation (Jiangsu Hanbon Science & Technology) was used. HPLC analysis was carried out on a Shimadzu Essentia LC-16 instrument. The system contained a DGU-20A$_{3R}$ degassing unit, an external column thermostat, an SPD-16 ultraviolet detector, an SIL-16 autosampler, two LC-16 pumps, and a Shimadzu workstation. MS spectra were obtained on a Waters QDa ESI mass spectrometer (Waters Corporation, Milford, MA, USA). NMR spectra were measured on a Bruker Avance 600 MHz (Bruker Corporation, Karlsruhe, Germany) with MeOH-d_4 as the solvent.

The column procured from Dr. Maisch & Co., Inc., Ammerbuch, Germany was a ReproSil-Pur C18 AQ (4.6 × 250 mm, 5 μm). XAmide (4.6 × 250 mm, 10 μm) and Click XION (4.6 × 250 mm, 5 μm) columns were obtained from Acchrom Technologies (Beijing, China). The Kromasil 60-5 Diol (4.6 × 250 mm, 5 μm) column used in this work was purchased from Kromasil (Bohus, Sweden). The Hypersil NH2 (4.6 × 250 mm, 5 μm) and silica (4.6 × 250 mm, 40–63 μm) columns were purchased from Dalian Elite Analytical Instrument (Liaoning, China).

Chromatographic-grade methanol, ethyl acetate, and dichloromethane used for HPLC were purchased from Shanghai Macklin Biochemical Co., Ltd. (Shanghai, China); analytical-grade dichloromethane and methanol were purchased from Yunnan Xinlanjing Chemical Industry, Qujing, China.

2.2. Sample Preparation

Roots of *Gymnadenia orchidis* were collected from Maixiu farm, Tongren county, Xining, China, in August 2017, and validated by professor Pengcheng Lin of the College of Pharmacy, Qinghai Nationalities University. A sample (No. 2017-L-3) was stored in the Qinghai Nationalities University Museum of Biology. *Tulasnellaceae* sp. was separated from the roots of *Gymnadenia orchidis*, and the identification of the *Tulasnellaceae* sp. is given in the File S1 of supporting information.

Tulasnellaceae sp. solid fermentation extracellular metabolites (32.5 g) were extracted thrice at room temperature with ethyl acetate (2.0 L and 2 h each). The combined extracted solution (6.0 L) was filtered and concentrated at 40 °C using a rotary evaporator to obtain the crude sample (2.2 g). The crude sample was dissolved in dichloromethane (6.0 mL) and then wet-loaded onto a silica gel medium-pressure column (40–63 μm, 49 mm × 460 mm, SiliCycle, Quebec, Canada). Methanol/dichloromethane were used as the eluent and the linear gradient was set to 0–30 min, 0–35% methanol, and 59.0 mL/min flow rate at room temperature. The chromatogram was recorded at 270 nm. After one separation, the collected eluate solution was concentrated and dried to finally obtain the target fraction (fraction 2, 586.0 mg).

2.3. *Chromatographic Conditions*

The crude sample and fraction 2 were analyzed on five representative analytical columns (Silica, XAmide, Hypersil NH_2, Kromasil 60-5 Diol, and Click XION). The crude sample (2.2 g) was dissolved in 6.0 mL of dichloromethane and passed through a 0.45 μm filter to obtain the sample solution (367.0 mg/mL). Fraction 2 (12.0 mg) was dissolved in 2.0 mL of methanol and passed through a 0.45 μm filter. The isocratic elution procedure was 100% dichloromethane for 16 min, the injection volume was 1.0 μL, and column temperature was maintained at room temperature. Chromatograms were recorded at 270 nm.

The fraction 2 solution (6.0 mg/mL) was analyzed on a ReproSil-Pur C18 AQ analytical column. The isocratic elution procedure was 95% methanol for 40 min, the injection volume was 1.0 μL, and the column temperature was maintained at room temperature.

The purity of fraction 2-2 was analyzed on a ReproSil-Pur C18 AQ analytical column, and the isocratic elution procedure was 100% methanol for 40 min with a 1 mL/min flow rate at room temperature. The monitoring wavelength was 270 nm.

3. Results and Discussion

3.1. *Extraction and Analysis of Endogenous Metabolites from Gymnadenia orchidis*

A total of 2.2 g of crude sample (6.8% yield) was obtained from 32.5 g of *Tulasnellaceae* sp. solid fermentation extracellular metabolites, which were then all dissolved in 6.0 mL of dichloromethane for subsequent analysis and pretreatment. The literature reveals that the structural characteristics and chemical properties of ergosterol give it a unique chromatographic retention behavior [25]. Therefore, the choice of chromatographic column filler is essential to obtain high-purity ergosterol from liquid chromatography.

To determine the optimal conditions, the chromatographic behavior of five representative chromatographic columns (Silica, XAmide, Hypersil NH_2, Kromasil 60-5 Diol, and Click XION) against the crude sample was investigated in normal phase mode. Isocratic elution of 100% dichloromethane was used on these columns with an injection volume of 1.0 μL at room temperature. The analysis result is shown in Figure 1. We observed that the crude sample showed up as two main fractions (fraction 1 and fraction 2) and exhibited good resolution on all five chromatographic columns (Figure 1A–E). Previous literature reports confirmed that ergosterol is the main component of the endogenous metabolite from *Gymnadenia orchidis*; therefore, fraction 2 was considered the target fraction for further purification. Medium-pressure liquid chromatography pretreatment of the crude extracts of natural products can remove some nontarget ingredients (natural pigments, macromolecular polymers, etc.), enrich target components, and simplify the workload of preparative high-pressure liquid chromatography. Silica gel is often used as filler for medium-pressure liquid chromatography pretreatment because of its low cost and high separation selectivity. As shown in Figure 1, all five chromatographic columns have the potential for pretreatment because of their effective separation of the crude sample. After comprehensive consideration of the quick peak flowing out, convenience, and moderate price, silica gel medium-pressure liquid chromatography was chosen for the pretreatment of the crude sample.

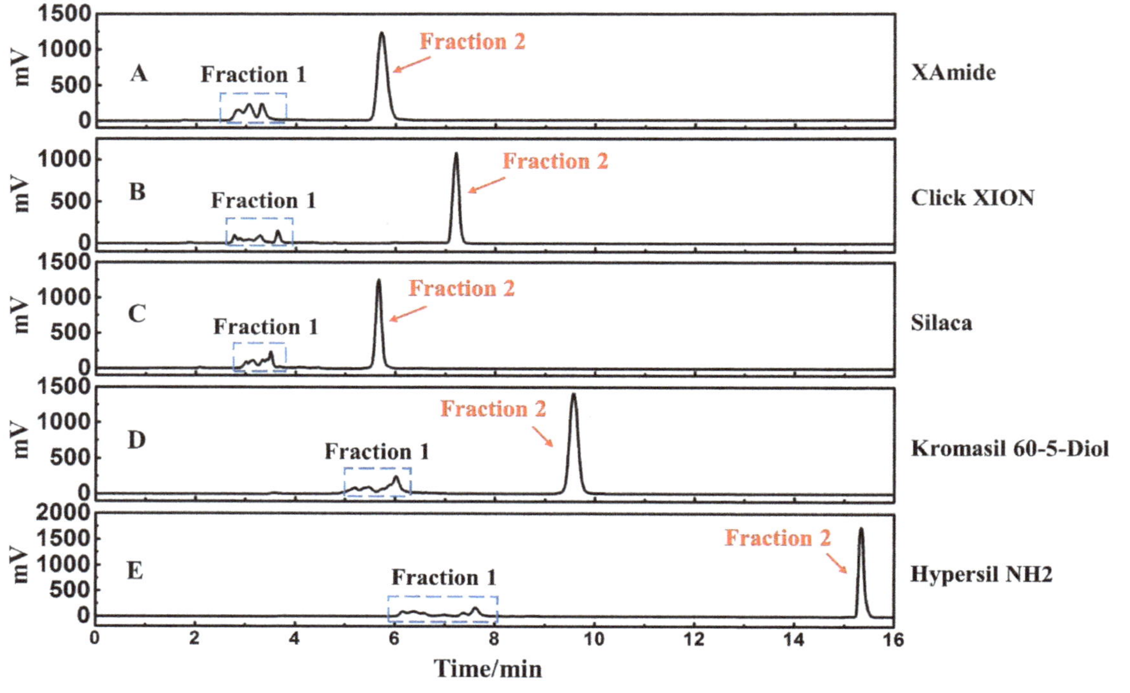

Figure 1. HPLC analysis of crude sample on five representative chromatography columns ((**A**): XAmide, (**B**): Click XION, (**C**): Silica, (**D**): Kromasil 60-5 Diol, and (**E**): Hypersil NH_2) under normal phase mode; injection volume: 1.0 μL. Conditions: isocratic elution: 0–16 min, 100% dichloromethane; monitoring wavelength: 270 nm; flow rate: 1.0 mL/min; column temperature: 30 °C.

3.2. Crude Sample Pretreatment via Medium-Pressure Chromatography

On the basis of the analysis results in Section 3.1, silica gel medium-pressure chromatography was used for the pretreatment of the crude sample. The supernatant liquid (6.0 mL) was removed from the top of the medium-pressure chromatographic tower, and the crude sample solution (6.0 mL) was added to the top of the medium-pressure chromatographic tower. Then, the medium-pressure chromatographic tower was connected to the preparative liquid chromatography for the pretreatment of the crude extract. The separation chromatogram is shown in Figure 2. Fraction 1 was nontarget ingredients, and fraction 2 was the target fraction for pretreatment. Due to the increase in the loading volume and the column diffusion effect, the retention time of the target component and peak shape were slightly different compared with the analytical chromatogram; however, it did not affect the recovery of fraction 2, and the visual separation of fraction 1 and fraction 2 was still achieved. After one separation, fraction 2 (586.0 mg) was obtained after concentration under reduced pressure, with a recovery of 26.6%. Fraction 2 (12.0 mg) was dissolved in 2.0 mL of methanol solution and passed through a 0.45 μm filter for subsequent analysis and separation.

3.3. Further Purification of Fraction 2 on ReproSil-Pur C18 AQ Chromatographic Column

Fraction 2 was analyzed on an XAmide and a ReproSil-Pur C18 AQ analytical chromatographic column. The result is shown in Figure 3. At the same time, the analysis results of the crude sample on the XAmide chromatographic column are also listed together for comparison (Figure 3B). Figure 3A,B demonstrates an excellent pretreatment effect; however, when the analysis was performed on a different selective column (Figure 3C), it was found that fraction 2 showed multiple low peaks (nontarget ingredients) and one main peak (target ingredient). We speculated that silica gel medium-pressure normal-phase liquid chromatography was used for the pretreatment of the crude sample, which has a natural separation selectivity different from the ReproSil-Pur C18 AQ column (reversed-phase chromatographic column). In normal phase mode, some components with consistent retention times of the target peaks cannot be visually separated; however, this problem is often well-solved in inverse phase mode because of the characteristic of selective variability [26–28]. Therefore, the ReproSil-Pur C18 AQ column was ultimately selected to further purify fraction 2.

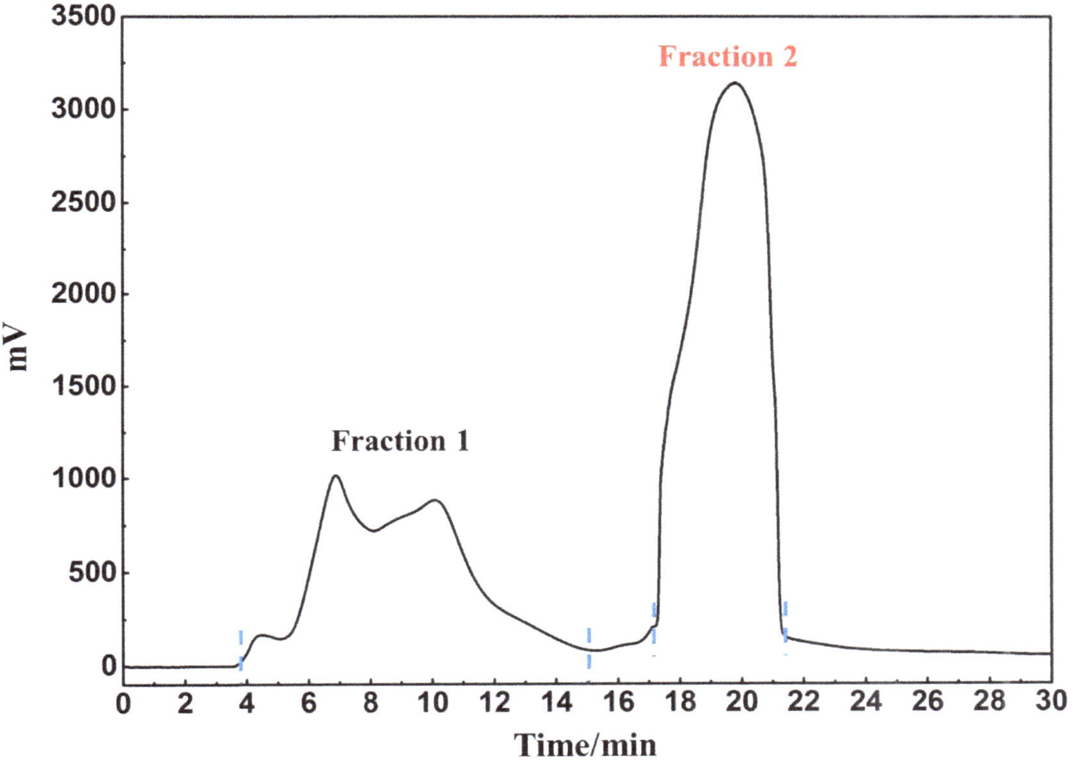

Figure 2. Pretreatment chromatogram of target components in *Tulasnellaceae* sp. with silica gel medium-pressure chromatographic tower; injection volume: 6.0 mL. Conditions: mobile phase A: dichloromethane, B: methanol; gradient: 0–30 min, 0–35% B; monitoring wavelength: 270 nm; flow rate: 59.0 mL/min.

3.4. Structural Characterization of Ergosterol

Fraction 2 (2.0 mL) solution was subjected to further isolation and purification. The loading volume for each preparation was 0.96 mL, and the eluate was concentrated under reduced pressure after 12 cycles of preparation. Figure 4A shows the comparison diagram for preparations, showing that the sample loading volume was large and reproducible and meets the needs of large-scale preparation. Finally, a total of 5.9 mg of compound was obtained with a recovery rate of 49.2%. The purity of the isolated compound was examined using a ReproSil-Pur C18 AQ analytical column. The chromatographic conditions are described in Section 2.3. As shown in Figure 4B, fraction 2-1 was obtained with >95% purity.

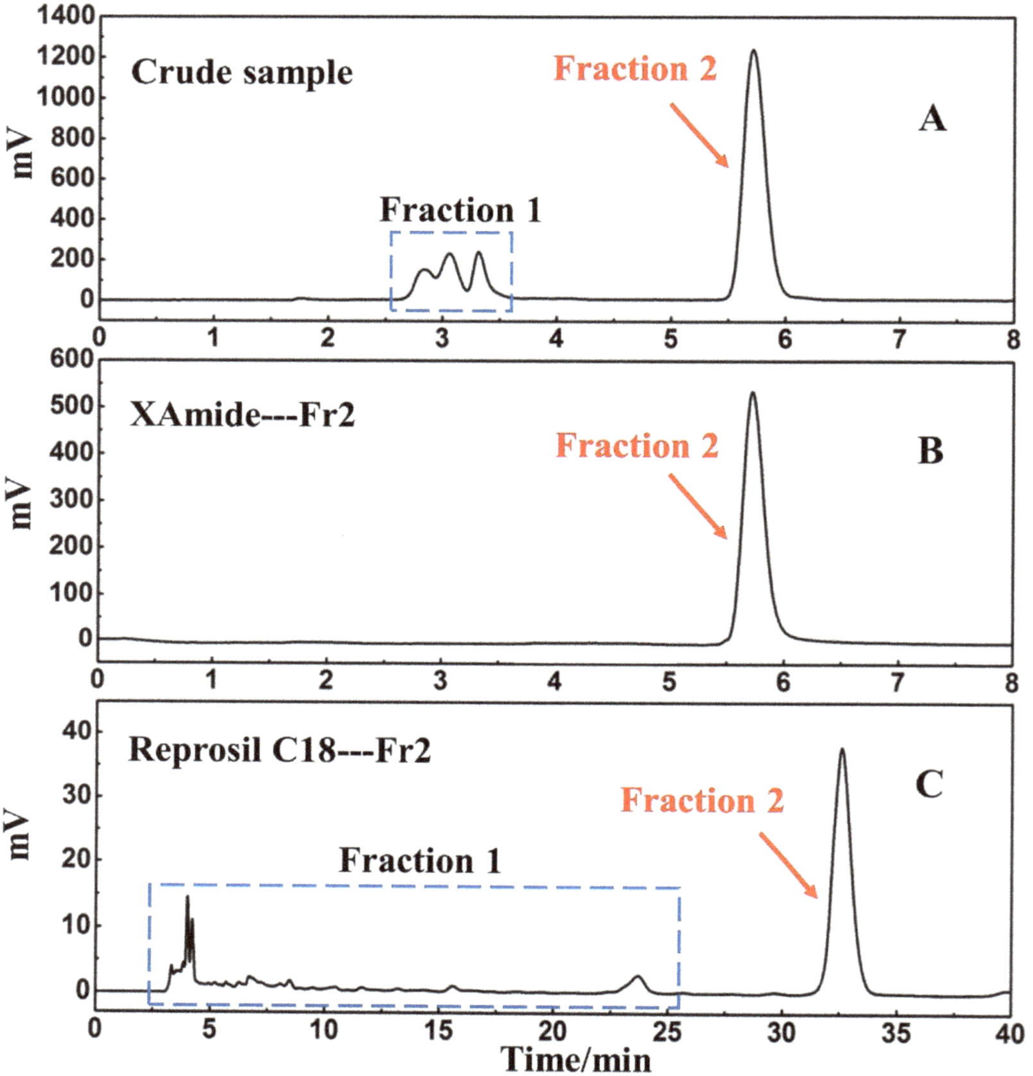

Figure 3. HPLC analysis of the target fraction on XAmide ((**A**): crude sample; (**B**): Fraction 2) and ReproSil C18 ((**C**): Fraction 2) analytical column; injection volume: 1.0 μL. Conditions of XAmide analytical column: isocratic elution: 0–16 min, 100% dichloromethane; monitoring wavelength: 270 nm; flow rate: 1.0 mL/min; column temperature: 30 °C. Conditions of ReproSil C18 analytical column: isocratic elution: 0–40 min, 100% methanol; monitoring wavelength: 270 nm; flow rate: 1.0 mL/min; column temperature: 30 °C.

To elucidate the structure of fraction 2-1, ESI-MS, ^{1}H NMR and ^{13}C NMR spectra were obtained and compared with published literature data. From the spectral data, we concluded that fraction 2-1 corresponded to ergosterol [6,29]. Original spectra are shown in the Figures S1–S3 of supporting information. The chemical structure of ergosterol is shown in Figure 4B.

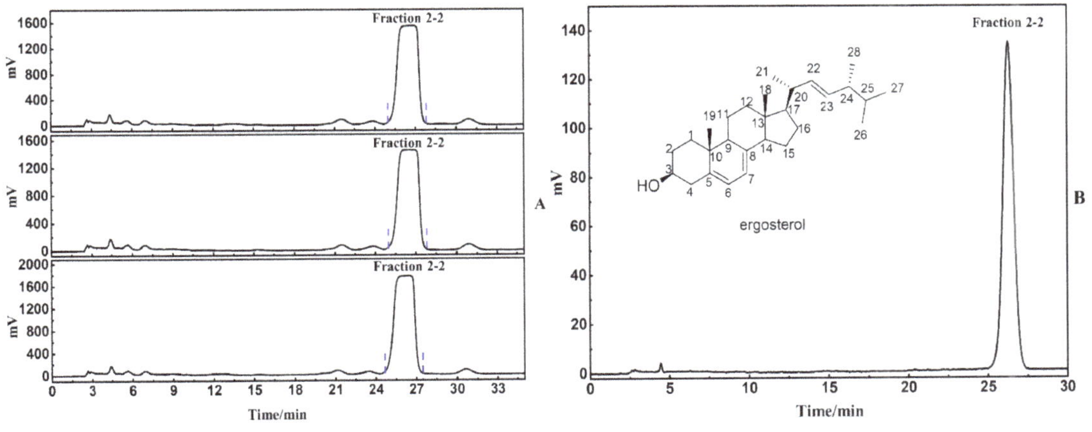

Figure 4. The preparative chromatogram and HPLC purity assay of the target fraction on ReproSil C18 analytical column in methanol/water (**A**). Conditions: mobile phase water and methanol; isocratic elution: 0–35 min, 100% methanol; monitoring wavelength: 270 nm; flow rate: 1.0 mL/min; injection volume of the preparative chromatogram: 80.0 µL; injection volume of HPLC purity assay: 1.0 µL; column temperature: 30 °C. The purity of the isolated compound was examined using a ReproSil-Pur C18 AQ analytical column (**B**). The isocratic elution procedure was 100% methanol for 40 min with a 1 mL/min flow rate at room temperature. The monitoring wavelength was 270 nm.

Fraction 2-2 (ergosterol, 5.9 mg, white powder, ESI-MS m/z: 398.34 $[M+2H]^+$): ^1H NMR (600 MHz, $CDCl_3$) 5.57 (1H, dd, *J* = 5.6, 2.5, H-6), 5.38 (1H, t, *J* = 5.4, H-7), 5.20 (1H, dd, *J* = 15.3, 7.5 Hz, H-22), 5.20 (1H, dd, *J* = 15.3, 7.5 Hz, H-23), 3.64 (1H, s, H-3), 1.179-2.502 (H-1,H-2, H-4, H-9,H-11,H-12, H-14, H-15, H-16, H-17, H-20, H-24, H-25), 1.04 (3H, d, *J* = 6.6, 21-Me), 0.95 (3H, s, 19-Me), 0.92 (3H, d, *J* = 6.9, 28-Me), 0.83 (6H, d, *J* = 6.9 Hz, 26-Me, 27-Me), 0.83 (6H, d, *J* = 6.9 Hz, 26-Me, 27-Me), 0.63 (3H, s, 18-Me); ^{13}C NMR (151 MHz, $CDCl_3$), 141.4 (C-8), 139.8 (C-5), 135.6 (C-22), 132.0 (C-23), 119.6 (C-6), 116.3 (C-7), 70.5 (C-3), 55.7 (C-17), 54.6 (C-14), 46.2 (C-9), 42.8 (C-13), 42.8 (C-24), 40.8 (C-4), 40.4 (C-20), 39.1 (C-12), 38.4 (C-1), 37.0 (C-10), 33.1 (C-25), 32.0 (C-2), 28.3 (C-16), 23.0 (C-15), 21.1 (C-11), 21.1 (C-21), 19.9 (C-26), 19.6 (C-27), 17.6 (C-28), 16.3 (C-19), 12.0 (C-18).

4. Conclusions

In this study, silica gel medium-pressure normal-phase liquid chromatography coupled with high-pressure reverse-phase liquid chromatography was developed for the large-scale purification of ergosterol from extracellular metabolites of *Tulasnellaceae* sp. First, the metabolite (2.2 g) of *Tulasnellaceae* sp. was pretreated by silica gel medium-pressure normal-phase chromatography to obtain fraction 2, and then the target compound was directly obtained after purification of fraction 2 using a ReproSil-Pur C18 AQ column, which was identified as ergosterol. Experiments showed that medium-pressure silica gel is well-suited for the pretreatment of natural products. Meanwhile, the medium-pressure normal-phase/high-pressure reverse-phase mode used in this study was able to separate ergosterol efficiently and on a large scale, which is expected to be further developed and applied in future methodological studies for the separation from natural products.

Supplementary Materials: The following supporting information can be downloaded at: https://www.mdpi.com/article/10.3390/separations9070176/s1, File S1: Species identification of the endophytic fungus *Tulasnellaceae* sp. From Gymnadenia orchidis, Figure S1: ESI mass spectrum of ergosterol, Figure S2: ^1H NMR spectrum (600 MHz) of ergosterol on ReproSil C18 analytical column in water/ethanol (in $CDCl_3$); Figure S3: ^{13}C NMR Spectrum (151 MHz) of ergosterol on ReproSil C18 analytical column in water/ethanol (in $CDCl_3$).

Author Contributions: Conceptualization, J.D. and P.L.; methodology, J.D. and Z.W.; formal analysis, Z.W. and Q.W.; resources, J.Z.; writing—original draft preparation, Z.W.; writing—review and editing, Z.W., J.D. and P.L. All authors have read and agreed to the published version of the manuscript.

Funding: This research was funded by the Qinghai Innovative Platform Project and Qinghai Medicinal Materials Resource Protection (the funding number was 2022-ZJ-Y14) and High Value Utilization Innovation Team (the funding number was 2021XJPI02).

Data Availability Statement: All data are available from the corresponding author upon request.

Conflicts of Interest: The authors declare no conflict of interest.

References

1. Purba, R.; Paenekoum, S.; Paeng, P. Development of a Simple High-Performance Liquid Chromatography-Based Method to Quantify Synergistic Compounds and Their Composition in Dried Leaf Extracts of *Piper Sarmentosum* Roxb. *Separations* **2021**, *8*, 152. [CrossRef]
2. Yang, B.B.; Li, S.; Zhang, R.P.; Wang, Y.; Shi, J.G. Quantitative analysis of four active constituents in Tibetan herb Gymnadenia conopsea by high-performance liquid chromatography. *Chin. J. Chin. Mater. Med.* **2009**, *34*, 1819–1822.
3. Wu, J.B.; Liu, C.P.; Lu, Y.B. Preparative separation of phytosterol analogues from green alga Chlorella vulgaris using recycling counter-current chromatography. *J. Sep. Sci.* **2017**, *40*, 2326–2334. [CrossRef]
4. Sha, Y.F.; Deng, C.H.; Zhang, H.B.; Xie, W.Y.; Liu, B.Z. Microwave-assisted silylation followed by gas chromatography/mass spectrometry for rapid determination of ergosterol in cigarettes. *J. Sep. Sci.* **2008**, *31*, 2451–2456. [CrossRef]
5. Chen, Y.H.; Xing, X.K.; Guo, S.X. The endophytic fungal community composition of *Gymnadenia conopsea* in Beijing. *Mycosystema* **2018**, *37*, 35–42.
6. Zhao, Y.; Xie, J.; Hou, S.; Ni, A.; Zhou, L. Isolation and characterization of ergosterol from monascus anka for anti-lipid peroxidation properties. *J. Mycol. Med.* **2020**, *30*, 101038.
7. Liu, Y.P.; Pu, C.J.; Wang, M.; He, J.; Li, Z.H.; Feng, T.; Xie, J.; Liu, J.K. Cytotoxic ergosterols from cultures of the basidiomycete Psathyrella candolleana. *Fitoterapia* **2019**, *138*, 104289. [CrossRef]
8. Yao, J.X.; Shi, Y.M.; Liu, Y.; He, S.; Ding, L.J.; Yang, F. Highly Oxidized Ergosterol Derivatives from the Fungus Nigrospora oryzae. *Chem. Nat. Compd.* **2019**, *55*, 390–392. [CrossRef]
9. Collins, M.D.; Shah, H.N.; Minnikin, D.E. A note on the separation of natural mixtures of bacterial menaquinones using phase thin-layer chromatography. *J. Appl. Bacteriol.* **2010**, *48*, 277–282. [CrossRef]
10. Xu, L.J.; Liu, S.B. Forecasting structure of natural products through color formation process by thin layer chromatography. *Food. Chem.* **2021**, *334*, 127496. [CrossRef] [PubMed]
11. Xu, M.J.; Qiao, Z.N.; Huang, G.S.; Long, M.F.; Yang, T.W.; Zhang, X.; Shao, M.L.; Xu, Z.H.; Rao, Z.M. Optimization of L-arginine Purification from *Corynebacterium crenatum* Fermentation Broth. *J. Sep. Sci.* **2020**, *43*, 2936–2948. [CrossRef] [PubMed]
12. Huang, Y.; Lei, C.W.; Song, H.; Yang, Z.Q.; Yuan, X.H.; He, X.S. Isolation and purification of ergosterol peroxide from *Xylaria Striata* by high-speed counter-current chromatography. *J. Food. Sci. Technol.* **2016**, *37*, 262–266.
13. Huang, L.Y.; Cao, Y.Y.; Xu, H.; Chen, G.N. Separation and purification of ergosterol and stigmasterol in *Anoectochilus roxburghii* (wall) Lindl by high-speed counter-current chromatography. *J. Sep. Sci.* **2011**, *34*, 385–392. [CrossRef]
14. Zhang, N.S.; Wang, J.B.; Wang, X.Y.; Wang, X.D.; Hu, F.L. Preparative isolation and purification of ergosterol from a strain of *Paecilomyces hepialid* by high-speed counter-current chromatography. *Se. Pu* **2010**, *28*, 68–72. [CrossRef]
15. Luca, C.D.; Lievore, G.; Bozza, D.; Buratti, A.; Catani, M. Downstream processing of therapeutic peptides by means of preparative liquid chromatography. *Molecules* **2021**, *26*, 4688. [CrossRef]
16. Perez, C.; Rani, M.; Phan, T. Optimization of high-performance liquid chromatography parameters for purification of oligonucleotide-a. *J. Anal. Chem.* **2022**, *13*, 39–50. [CrossRef]
17. Dang, J.; Du, Y.R.; Wang, Q.; Dawa, Y.Z.; Chen, C.B.; Wang, Q.L.; Ma, J.B.; Tao, Y.D. Preparative isolation of arylbutanoid-type phenol [(-)-rhododendrin] with peak tailing on conventional C18 column using middle chromatogram isolated gel column coupled with reversed-phase liquid chromatography. *J. Sep. Sci.* **2020**, *23*, 3233–3241. [CrossRef]
18. Bitterling, H.; Schaefer, U.; Krammer, G.E.; Meier, L.; Steingass, C.B. Investigations into the natural occurrence of 1-phenylethyl acetate (styrallyl acetate). *J. Agric. Food Chem.* **2020**, *68*, 8613–8620. [CrossRef]
19. Angelis, A.; Michailidis, D.; Antoniadi, L.; Stathopoulos, P.; Skaltsounis, L. Pilot continuous centrifugal liquid-liquid extraction of extra virgin olive oil biophenols and gram-scale recovery of pure oleocanthal, oleacein, mfoa, mfla and hydroxytyrosol. *A. Sep. Purif. Technol.* **2021**, *255*, 117692. [CrossRef]
20. Pan, G.Q.; Shen, J.W.; Ma, Y.H.; He, Y.F.; Bao, Y.; Li, R.R.; Wang, S.S.; Wang, Q.; Lin, P.C.; Dang, J. Preparative separation of isoquinoline alkaloids from *Corydalis impatiens* using a middle-pressure chromatogram isolated gel column coupled with two-dimensional liquid chromatography. *J. Sep. Sci.* **2019**, *42*, 3182–3190. [CrossRef]
21. Dawa, Y.; Du, Y.; Wang, Q.; Chen, C.; Zou, D.; Qi, D.; Ma, J.; Dang, J. Targeted isolation of 1,1-diphenyl-2-picrylhydrazyl inhibitors from saxifraga atrata using medium- and high- pressure liquid chromatography combined with online high performance liquid chromatography–1,1-diphenyl-2- picrylhydrazyl detection. *J. Chromatogr. A* **2021**, *1635*, 461690. [CrossRef] [PubMed]

22. Park, S.I.; Park, S.C.; Kim, S.R.; Jang, Y.P. Two-step purification method for aging pigments a2e and iso-a2e using medium pressure liquid chromatography. *B Korean Chem. Soc.* **2016**, *37*, 1541–1544. [CrossRef]
23. Zheng, X.; Qin, Y.; Meng, X.; Jin, Z.; Wang, J. Synthesis of polyethylene glycol functional bonded silica gel for selective recognition and separation of α-cyclodextrin. *J. Chromatogr. A* **2021**, *1639*, 461917. [CrossRef]
24. Gunawan, S.; Pamungkas, B.; Primaswari, C.S.; Hapsari, S.; Aparamarta, H.W. Calophyllolide separation from calophyllum inophyllum oil by silica gel adsorption. *Mater. Sci. Forum.* **2020**, *988*, 101–107.
25. Ying, Z.; Ouyang, X.; Chen, J.; Zhao, L.; Qiu, X. Separation of aromatic monomers from oxidatively depolymerized products of lignin by combining sephadex and silica gel column chromatography. *Sep. Purif. Technol.* **2018**, *191*, 250–261.
26. Xiu, W.; Yupei, Z.; Nan, W.; Jingya, C.; Yanduo, T.; Ruitao, Y. A Method to Separate Two Main Antioxidants from Lepidium latifolium L. Extracts Using Online Medium Pressure Chromatography Tower and Two-Dimensional Inversion/Hydrophobic Interaction Chromatography Based on Online HPLC-DPPH Assay. *Separations* **2021**, *12*, 238.
27. Zhou, W.; Liu, Y.; Wang, J.; Guo, Z.; Shen, A.; Liu, Y.; Liang, X. Application of two-dimensional liquid chromatography in the separation of traditional Chinese medicine. *J. Sep. Sci.* **2020**, *43*, 87–104. [CrossRef]
28. Fan, Y.; Fu, Y.; Fu, Q.; Cai, J.; Xin, H.; Dai, M.; Jin, Y. Purification of flavonoids from licorice using an off-line preparative two-dimensional normal-phase liquid chromatography/reversed-phase liquid chromatography method. *J. Sep. Sci.* **2016**, *39*, 2710–2719. [CrossRef]
29. Nowak, R.; Drozd, M.; Mendyk, E.; Lemieszek, M.; Krakowiak, O.; Kisiel, W.; Rzeski, W.; Szewczyk, K. A new method for the isolation of ergosterol and peroxyergosterol as active compounds of hygrophoropsis aurantiaca and in vitro antiproliferative activity of isolated ergosterol peroxide. *Molecules* **2016**, *21*, 946. [CrossRef]

separations

MDPI

Article

Simultaneous Determination of Methyl Nicotinate and Three Salicylic Acid Derivatives in Pain Relief Spray Using HPLC–DAD

Hazim M. Ali

Department of Chemistry, College of Science, Jouf University, P.O. Box 2014, Sakaka 72388, Aljouf, Saudi Arabia; hmali@ju.edu.sa; Tel.: +966-53-7107-043

Abstract: For the first time, the high-performance liquid chromatography–diode array detector (HPLC–DAD) approach was operated for the simultaneous assessment of methyl nicotinate (MN), methyl salicylate (MS), ethyl salicylate (ES) and 2-hydroxyethyl salicylate (HES) in one pharmaceutical formulation. The limits of detection of MN, HES, MS and ES were found to be 0.0144, 0.0455, 0.0087 and 0.0061 μg/mL. The recovery percentages and relative standard deviations ranged from 93.48 to 102.12% and 0.301 to 6.341% for all active ingredients. Accordingly, the previously described data demonstrate the sensitivity, accuracy and precision of the developed method. Therefore, the investigated approach was effectively applied for the simultaneous assessment of MN, HES, MS and ES in DEEP HEAT Spray.

Keywords: methyl nicotinate; methyl salicylate; ethyl salicylate; 2-hydroxyethyl salicylate; pain relief spray

Citation: Ali, H.M. Simultaneous Determination of Methyl Nicotinate and Three Salicylic Acid Derivatives in Pain Relief Spray Using HPLC–DAD. *Separations* **2022**, *9*, 93. https://doi.org/10.3390/separations9040093

Academic Editor: Victoria Samanidou

Received: 18 March 2022
Accepted: 5 April 2022
Published: 6 April 2022

1. Introduction

Esters of salicylic acid such as methyl salicylate, ethyl salicylate and 2-hydroxyethyl salicylate (Figure 1) are used as analgesic and rubefacient in many topical creams and sprays for the relief of muscle and joint pain [1–7]. Methyl nicotinate is methyl ester of nicotinic acid (Figure 1), has a vasodilator property, enhances the topical penetration of active ingredients in cream and sprays and also has an effective role for relief of pain and aches in joints, tendons and muscles [8,9].

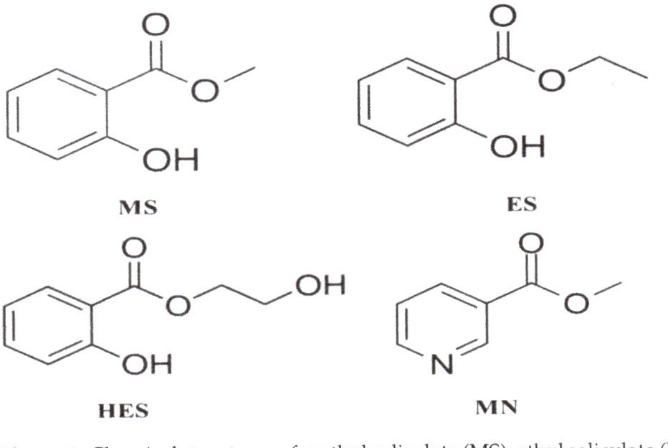

Figure 1. Chemical structures of methyl salicylate (MS), ethyl salicylate (ES), 2-hydroxyethyl salicylate (HES) and methyl nicotinate (MN).

Therefore, the presence of methyl salicylate, ethyl salicylate and 2-hydroxyethyl salicylate and methyl nicotinate in one formulation enhances their efficiency for pain relief [2].

To the best of our knowledge, the literature contains a few methods for the individual determination of methyl salicylate, ethyl salicylate and 2-hydroxyethyl salicylate and methyl nicotinate in different samples. High performance liquid chromatography (HPLC) was used for the individual assessment of methyl salicylate and methyl nicotinate in pharmaceutical formulations and medicinal plants [2,8,10,11], while gas chromatography mass spectrometry (GC–MS) was utilized to estimate methyl salicylate and ethyl salicylate in biological fluids [2,12].

Only one research paper published by Pauwels et al. [2] focuses on the determination of methyl salicylate, ethyl salicylate and 2-hydroxyethyl salicylate in one topical formulation by using two chromatographic techniques. The two methods used by Pauwels et al. [2] included the use of a gas chromatography flame ionization detector (GC-FID), which was applied for the simultaneous assessment of methyl salicylate and ethyl salicylate, while a liquid chromatography ultraviolet detector (LC-UV) was used for the determination of 2-hydroxyethyl salicylate.

Therefore, the proposed work presents the first approach for the simultaneous assessment of methyl salicylate, ethyl salicylate and 2-hydroxyethyl salicylate and methyl nicotinate in one tropical formulation based on the high-performance liquid chromatography supplied with a diode array detector (HPLC–DAD). Besides, the separation efficiency, simplicity, sensitivity, reliability and total analysis time of the investigated approach for the assessment of the four analytes will be evaluated for use in quality control protocol as well as pharmacokinetic studies.

2. Materials and Methods

2.1. Instrument

Thermo Scientific Dionex UltiMate 3000 UHPLC connected to DAD-3000 diode array detector was adjusted for the assessment of MN, HES, SA, MS and ES in solutions. The analysis data were recorded via a Chromeleon™ 7.2 Chromatography Data System. A hypersil GOLD column (250 mm length, 4 mm inner diameter, 5 µm particle size (Thermo Scientific, Waltham, MA, USA) was used for the separation of analytes.

2.2. Chemicals and Materials

MN, HES, MS, ES, salicylic acid (SA), acetonitrile, formic acid and methanol were purchased from Sigma-Aldrich (Steinheim, Germany). A Barnstead™ Smart2Pure™ water purification system was used for deionized water production.

DEEP HEAT Spray (150 mL) contains 1.6% of MN, 5% HES, 1% of MS and 5% of ES, its manufacturer and marketing authorization holder is the Mentholatum Co. Ltd., East Kilbride, Scotland, UK, and it was purchased from local community pharmacies in Saudi Arabia.

2.3. Preparation of Stock Solutions

MN, HES, MS, ES and SA (as internal standard) stock solutions were made in 50 mL of HPLC grade methanol at a concentration of 1000 µg/mL and diluted to the needed concentration using the same solvent to prepare working solutions.

2.4. Preparation of Spray Solution

In total, 2.0 mL of spray content were transferred to 50 mL glass flask and diluted to the mark by HPLC grade methanol. MN, HES, MS and ES working solutions of 14.57, 45.55, 9.11 and 45.55 µg/mL, respectively, in presence of 0.5 µg/mL of SA internal standard, were prepared after a series of dilutions for the prepared spray solution.

2.5. Chromatographic Conditions

The separation of MN, HES, SA, MS and ES was achieved through a Hypersil GOLD column (250 mm length, 4 mm inner diameter, 5 µm particles size (Thermo Scientific, Waltham, MA, USA). The composition of mobile phase was 50% methanol: 50% acetonitrile (A) and water acidified with formic acid (0.1%) (B) in the volume percent of 70:30 (*v/v*) at a flow rate of 0.5 mL/min with an isocratic elution mode. The mixture components were detected at a wavelength of 210 nm and at room temperature (25 °C). The injection volume of standard solutions and samples was 10 µL.

2.6. Validation of Assay Approach

The precision, accuracy, limit of detection (LOD), limit of quantification (LOQ), linearity and system suitability variables were used to validate the rapid identification of MN, HES, MS and ES using the HPLC approach. The linearity of developed method was tested by using a series of concentration levels of MN, HES, MS and ES standards ranging from 0.03–100, 0.05–50, 0.03–50 and 0.03–50 µg/mL, respectively, in presence of SA as internal standard. Accuracy of the method was calculated for the five concentration levels (0.07, 0.5, 1, 5, 20 µg/mL) of MN, HES, MS and ES in triplicate by using the following formula: recovery % = determined value/added value × 100. Intra-day and inter-day precisions were evaluated for the mentioned above concentration levels of each component in triplicate by estimating the relative standard deviation (RSD%) = (σ/mean determine concentration) × 100, where σ is a standard deviation of intercept). Limit of quantification (LOQ), and limit of detection (LOD) of MN, HES, MS and ES were calculated from linear regression equations dependent on the slope and standard deviation of the intercept via applying the following formula: LOD = 3 σ/S and LOQ = 10 σ/S, where σ is the standard deviation of intercept and S is the slope of the calibration curve. The system suitability was evaluated by calculating selectivity factor (α), resolution (R_s), capacity factor (K′), column efficiency (N), tailing factor (T) and height equivalent to theoretical plate (HETP).

3. Result

3.1. Development and Optimization Processes

Several chromatographic experiments were tested to obtain the HPLC chromatograms with the best separation and resolution of MN, HES, MS and ES peaks in a short time of analysis, e.g., mobile phase composition, elution mode, rate of flow, kind of column, temperature of column and recognition wavelength. Three distinct columns, including Thermo Scientific ACCLAIM™ 120 C8 (4.6 × 150 mm, 5 µm), Hypersil GOLD (4 × 250 mm, 5 µm) and ACCLAIM™ 120 C18 (4.6 × 150 mm, 5 µm), were tested at different temperatures. Finally, the Hypersil GOLD (4 × 250 mm, 5 µm) column and 25 °C were found to be the best column and temperature for the separation and determination of the MN, HES, MS and ES mixtures. Furthermore, the mobile phase compositions were examined using a variety of solvents (methanol and acetonitrile), acids (formic acid, acetic acid and phosphoric acid) and buffers as well as elution mode and flow rate. The suitable separation and resolution were achieved by the isocratic elution mode using 50% methanol and 50% acetonitrile (A) and water acidified with formic acid (0.1%) (B) in the volume percent of 70:30 (*v/v*) at a flow rate of 0.5 mL/min with an isocratic elution mode. On the other hand, the UV detector was adjusted at 210 nm in order to get the desired sensitivity for MN, HES, MS and ES. Under these conditions, separation of MN, HES, MS and ES was carried out in 11 min and retention times were 5.57, 6.03. 8.09, and 10.04 min, respectively, as shown in Figures 2a and S1. The good separation of MN, HES and SA was evidenced by focusing the time range on the three components only as displayed in Figure 2b.

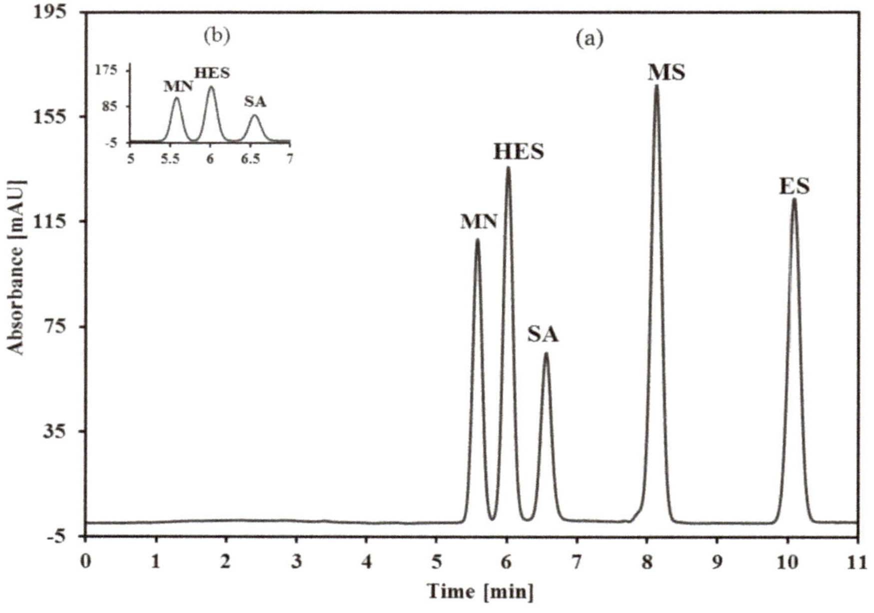

Figure 2. Chromatogram for MN, HES, SA, MS and ES at optimum conditions. (**a**) complete chromatogram; (**b**) the enlarged view of the scale range from 5 to 7 min.

3.2. Method Validation

The ICH recommendations [13] were applied for validation for the efficiency of the HPLC approach.

3.2.1. Linearity and Calibration Curve

The linear range and calibration curve equation of MN, HES, MS and ES were described in Table 1. Correlation coefficient values exceeded 0.99, as listed in Table 1.

Table 1. The results of regression equations, LOD and LOQ.

Analyte	Regression Equation	R	Linear Range (μg/mL)	LOD (μg/mL)	LOQ (μg/mL)
MN	Y = 0.968 x − 0.713	0.999	0.03–100	0.0144	0.0478
HES	Y = 2.485 x + 0.315	0.996	0.05–50	0.0455	0.1516
MS	Y = 3.151 x + 0.083	0.998	0.03–50	0.0087	0.0289
ES	Y = 2.926 x + 0.579	0.998	0.03–50	0.0061	0.0204

3.2.2. LOD and LOQ Assessment

The sensitivity of the investigated HPLC method toward MN, HES, MS and ES was confirmed by the calculation of LOD and LOQ. According to the formulas mentioned above, the LOD and LOQ of MN, HES, MS and ES were depicted in Table 1. The LOD and LOQ values of each component refer to the great sensitivity of the investigated approach when compared with the reported methods [2,8,10–12].

3.2.3. Accuracy and Precision

Accuracy and precision of the suggested HPLC approach for the assessment of MN, HES, MS and ES were calculated according to the formulas mentioned above by testing five concentration levels of each component (0.07, 0.5, 1, 5 and 20 μg/mL) for three replicates as depicted in Table 2. The recovery (%) and relative standard deviation (RSD%) were found in the range from 93.48 to 102.12% and 0.301 to 6.341%, respectively. The obtained results of intra- and inter-day assays were within the accepted limits.

Table 2. Accuracy and precision of the proposed approach for the assessment of MN, HES, MS and ES.

Analytes	Conc. Added (μg/mL)	Intra-Day			Inter-Day		
		Conc. Found (μg/mL) ± SD	Recovery (%)	RSD (%)	Conc. Found (μg/mL) ± SD	Recovery (%)	RSD (%)
MN	0.07	0.068 ± 0.002	97.25	2.672	0.067 ± 0.003	95.38	3.979
	0.5	0.510 ± 0.008	101.91	1.653	0.496 ± 0.011	99.22	2.115
	1	1.007 ± 0.059	100.68	5.886	1.006 ± 0.035	100.62	3.519
	5	5.099 ± 0.194	101.97	3.810	5.018 ± 0.205	100.36	4.082
	20	19.853 ± 0.073	99.263	0.369	19.85 ± 0.061	99.23	0.301
HES	0.07	0.068 ± 0.001	97.29	1.380	0.069 ± 0.004	97.95	6.341
	0.5	0.508 ± 0.013	101.68	2.532	0.503 ± 0.023	100.59	4.571
	1	0.982 ± 0.049	98.180	5.028	1.012 ± 0.033	101.21	3.218
	5	4.80 ± 0.091	96.08	1.890	4.719 ± 0.211	94.38	4.467
	20	20.06 ± 0.180	100.28	0.897	19.614 ± 0.709	98.07	3.616
MS	0.07	0.070 ± 0.001	100.50	1.421	0.069 ± 0.002	98.73	2.794
	0.5	0.503 ± 0.007	100.61	1.445	0.473 ± 0.008	94.58	1.782
	1	1.021 ± 0.011	102.12	1.121	0.943 ± 0.011	94.27	1.129
	5	4.915 ± 0.034	98.29	0.683	4.674 ± 0.041	93.48	0.861
	20	19.982 ± 0.086	99.91	0.429	20.082 ± 0.092	100.41	0.461
ES	0.07	0.068 ± 0.002	97.02	2.949	0.067 ± 0.002	96.23	2.786
	0.5	0.510 ± 0.008	102.05	1.637	0.492 ± 0.005	98.47	0.971
	1	0.993 ± 0.018	99.33	1.838	1.008 ± 0.012	100.76	1.191
	5	7.828 ± 0.076	96.56	1.580	4.871 ± 0.046	97.39	0.951
	20	19.786 ± 0.154	98.93	0.776	19.871 ± 0.348	99.36	1.749

3.2.4. System Suitability Testing (SST)

Several SST parameters were measured including selectivity factor (α), resolution (Rs), capacity factor (K′), column efficiency (N), tailing factor (T) and height equivalent to theoretical plate (HETP) to check and ensure ongoing HPLC system performance for the simultaneous determination of MN, HES, MS and ES. As depicted in Table 3, the values of Rs, α, T, K′, N, HETP ranged from 3.195 to 15.96, 1.16 to 1.41, 0.941 to 1.03, 1.05 to 2.50, 8543 to 17918 and 0.00008 to 0. 002, respectively. These values were found to be within the recommended limits [7], suggesting the accessibility and efficiency of the investigated HPLC approach for the determination of the four analytes.

Table 3. System suitability testing parameters of the proposed HPLC method.

Parameters	Obtained Value				Reference Value [13]
	MN	HES	MS	ES	
Resolution (R_s)	3.195	15.96	12.125	9.60	<1.5
Selectivity factor (α)	1.16	1.17	1.41	1.37	< 1
Tailing factor(T)	0.941	0.947	1.00	1.03	>1.5–2 or >2
Capacity factor(K′)	1.05	1.30	1.82	2.50	1–10 acceptable
Column efficiency (n)	8543	13877	17918	17479	Increase with efficiency of the separation
HETP [b]	0.002	0.001	0.0008	0.00008	The smaller the value the higher the column efficiency

HETP [b] = height equivalent to theoretical plate, (cm/plate).

3.3. Application of the Method

The investigated HPLC–DAD approach was successfully operated for the simultaneous assessment of MN, HES, MS and ES in DEEP HEAT Spray. The values of the recovery percentage of MN, HES, MS and ES, ranged from 92.04% to 101.14% with the standard deviation not exceeding 0.56% as depicted in Table 4, support this point.

Table 4. Analysis of MN, HES, MS and ES in Deep Heat Spray by the proposed HPLC method.

Component	Taken (μg/mL)	Recovery %
MN	14.57	97.88 ± 0.01
HES	45.55	92.04 ± 0.56
MS	9.11	101.14 ± 0.13
ES	45.55	94.39 ± 0.40

4. Conclusions

For the first time, an unsophisticated, dependable, accurate and precise HPLC approach was established for the simultaneous determination of methyl nicotinate, methyl salicylate, ethyl salicylate and 2-hydroxyethyl salicylate in one formulation. In addition, the investigated method has the advantage of eluting the four analytes in a short analytical run time. The recovery percentages and relative standard deviations ranged from 93.48 to 102.12% and 0.301 to 6.341% for all analytes. As a result, the proposed quantitative approach can be used successfully for quality control laboratories and routine analysis of the methyl nicotinate, methyl salicylate, ethyl salicylate and 2-hydroxyethyl salicylate.

Supplementary Materials: The following supporting information can be downloaded at: https://www.mdpi.com/article/10.3390/separations9040093/s1, Figure S1: Chromatogram for MN, HES, SA, MS and ES at optimum conditions (by using methanol as dilution solvent and before use mobile phase as diluent).

Funding: This work was funded by the Deanship of Scientific Research at Jouf University under grant No. (DSR-2021-03-0313).

Institutional Review Board Statement: Not applicable.

Informed Consent Statement: Not applicable.

Data Availability Statement: Not applicable.

Acknowledgments: The author extends his appreciation to the Deanship of Scientific Research at Jouf University for funding this work through research grant No (DSR-2021-03-0213). The author also thanks Hassan M. A. Hassan and Mohammed Gamal for their kind advice and support.

Conflicts of Interest: The author does not have any conflict of interest to declare related to this work.

References

1. Mason, L.; Moore, R.A.; Derry, S.; Edwards, J.E.; McQuay, H.J. Systematic Review of Topical Capsaicin for the Treatment of Chronic Pain. *BMJ* **2004**, *328*, 991. [CrossRef] [PubMed]
2. Pauwels, J.; D'Autry, W.; Van den Bossche, L.; Dewever, C.; Forier, M.; Vandenwaeyenberg, S.; Wolfs, K.; Hoogmartens, J.; Van Schepdael, A.; Adams, E. Optimization and Validation of Liquid Chromatography and Headspace-Gas Chromatography Based Methods for the Quantitative Determination of Capsaicinoids, Salicylic Acid, Glycol Monosalicylate, Methyl Salicylate, Ethyl Salicylate, Camphor and l-Menthol in A. *J. Pharm. Biomed. Anal.* **2012**, *60*, 51–58. [CrossRef] [PubMed]
3. Rainsford, K.D.; Whitehouse, M.W. Anti-Inflammatory/Anti-Pyretic Salicylic Acid Esters with Low Gastric Ulcerogenic Activity. *Agents Actions* **1980**, *10*, 451–456. [CrossRef] [PubMed]
4. Miles, S. Methyl Salicylate. In *xPharm: The Comprehensive Pharmacology Reference*; Enna, S.J., Bylund, D.B., Eds.; Elsevier: New York, NY, USA, 2007; pp. 1–6. ISBN 978-0-08-055232-3.
5. Michel, P.; Granica, S.; Magiera, A.; Rosińska, K.; Jurek, M.; Poraj, Ł.; Olszewska, M.A. Salicylate and Procyanidin-Rich Stem Extracts of Gaultheria Procumbens L. Inhibit Pro-Inflammatory Enzymes and Suppress Pro-Inflammatory and Pro-Oxidant Functions of Human Neutrophils Ex Vivo. *Int. J. Mol. Sci.* **2019**, *20*, 1753. [CrossRef] [PubMed]
6. Horak, J.; Hemmer, W.; Focke, M.; Götz, M.; Jarisch, R. Contact Dermatitis from Anti-Inflammatory Gel Containing Hydroxyethyl Salicylate. *Contact Dermat.* **2002**, *47*, 109–125. [CrossRef] [PubMed]
7. Kučera, M.; Kolar, P.; Barna, M.; Kučera, A.; Hladiková, M. Arnica/Hydroxyethyl Salicylate Combination Spray for Ankle Distortion: A Four-Arm Randomised Double-Blind Study. *Pain Res. Treat.* **2011**, *2011*, 365625. [CrossRef] [PubMed]
8. De Spiegeleer, B.; Van den Bossche, W.; De Moerloose, P.; Stevens, H. Determination of Methyl Nicotinate in Pharmaceutical Creams by High-Performance Thin-Layer Chromatography. *Chromatographia* **1985**, *20*, 249–252. [CrossRef]
9. Jumbelic, L.C.; Liebel, F.T.; Southall, M.D. Establishing a Minimal Erythema Concentration of Methyl Nicotinate for Optimum Evaluation of Anti-Inflammatories. *Skin Pharmacol. Physiol.* **2006**, *19*, 147–152. [CrossRef] [PubMed]
10. Abounassif, M.A.; Abdel-Moety, E.M.; Gad-Kariem, R.A. HPLC-Quantification of Diethylamine Salicylate and Methyl Nicotinate in Ointments. *J. Liq. Chromatogr. Relat. Technol.* **1992**, *15*, 625–636. [CrossRef]
11. Parker, D.; Martinez, C.; Stanley, C.; Simmons, J.; McIntyre, I.M. The Analysis of Methyl Salicylate and Salicylic Acid from Chinese Herbal Medicine Ingestion. *J. Anal. Toxicol.* **2004**, *28*, 214–216. [CrossRef] [PubMed]
12. Kakkar, T.; Mayersohn, M. Simultaneous Quantitative Analysis of Methyl Salicylate, Ethyl Salicylate and Salicylic Acid from Biological Fluids Using Gas Chromatography–Mass Spectrometry. *J. Chromatogr. B Biomed. Sci. Appl.* **1998**, *718*, 69–75. [CrossRef]
13. Center for Drug Evaluation and Research, U.S. *Food and Drug Administration, FDA. Reviewer Guidance, Validation of Chromatographic Methods*; FDA: Silver Spring, MD, USA, 1994.

separations

MDPI

Article

Analysis of Volatiles in *Senecio anteuphorbium* Essential Oil with a Focus on Its Allelopathic Effect by Means of Gas Chromatography

Soukaina Ouhaddou [1], Abdellah Aghraz [2], Widad Ben Bakrim [2,3], Saida Sissi [1], Mustapha Larhsini [1], Mohamed Markouk [1], Khalid Bekkouche [1], Sara Arrigo [4,5], Nicola Cicero [4,5,6], Rosaria Costa [4,*] and Rossella Vadalà [4]

1 Laboratory of Agri-Food, Biotechnologies and Valorization of Plant Bioresources (AGROBIOVAL), Team of Protection and Valorization of Plant Resources, Faculty of Sciences Semlalia, BP: 2390, Cadi Ayyad University, Marrakesh 40000, Morocco; Ouhaddousoukaina@gmail.com (S.O.); saida20sissi@gmail.com (S.S.); larhsini@uca.ma (M.L.); markouk@uca.ac.ma (M.M.); Bekkouche@uca.ma (K.B.)

2 Agrobiosciences Research Department, Mohammed VI Polytechnic University (UM6P), Lot 660–Hay Moulay Rachid, Ben-Guerir 43150, Morocco; abdellah.aghraz@gmail.com (A.A.); Widadbenbakrim@gmail.com (W.B.B.)

3 African Sustainable Agriculture Research Institute (ASARI), Mohammed VI Polytechnic University (UM6P), Laayoune 70040, Morocco

4 Department of Biomedical, Dental, Morphological and Functional Imaging Sciences (BIOMORF), University of Messina, Viale Annunziata, 98168 Messina, Italy; saraarrigo@hotmail.it (S.A.); nicola.cicero@unime.it (N.C.); rossella.vadala@unime.it (R.V.)

5 Science4Life s.r.l., A Spin-Off of the University of Messina, Viale Annunziata, 98168 Messina, Italy

6 Consorzio di Ricerca sul Rischio Biologico in Agricoltura CO.RI.BI.A. c/o Istituto Zooprofilattico Sperimentale della Sicilia, Via Gino Marinuzzi, 3, 90129 Palermo, Italy

* Correspondence: costar@unime.it

Citation: Ouhaddou, S.; Aghraz, A.; Ben Bakrim, W.; Sissi, S.; Larhsini, M.; Markouk, M.; Bekkouche, K.; Arrigo, S.; Cicero, N.; Costa, R.; et al. Analysis of Volatiles in *Senecio anteuphorbium* Essential Oil with a Focus on Its Allelopathic Effect by Means of Gas Chromatography. *Separations* **2022**, *9*, 36. https://doi.org/10.3390/separations9020036

Academic Editor: Grzegorz Boczkaj

Received: 16 December 2021
Accepted: 25 January 2022
Published: 29 January 2022

Abstract: The present study aimed to investigate *Senecio anteuphorbium*, an endemic plant growing in West Morocco and widely used in local folk medicine. The essential oil (EO) extracted from the aerial parts was analyzed by gas chromatography and tested for allelopathic activity. The quantitation of the volatiles was carried out by means of GC-FID with response factors, which were validated through reliable calibration procedures, based on external and internal standardization. This analytical approach allowed to define the real concentration of each constituent (weight%, g/100 g) alongside the conventional relative percent. On the other hand, the identification process was supported by a dual matching based on both mass spectra and retention indices. The essential oil resulted in being rich in sesquiterpenes, with the predominant constituents being bicyclogermacrene (22.75 g/100 g), spathulenol (25.26 g/100 g), epi-γ-eudesmol (6.8 g/100 g), and selina-4,11-diene (5.08 g/100 g). The allelopathic effect was evaluated by studying the inhibition of the germination and growth of *Lactuca sativa* seeds. A potent allelopathic effect was recorded by the essential oil at a dose of 0.281 mg/mL, with almost a total inhibition of germination.

Keywords: *Senecio anteuphorbium*; response factors; true quantitation; allelopathy

1. Introduction

Nowadays, the use of natural resources has attracted the attention of many researchers due to their diversity, durability, bioactivity, and eco-friendly characteristics. Essential oils (EOs) are one of the richest sources of bioactive metabolites that justify their numerous biological activities, such as antibacterial, antiviral, insecticidal, and allelopathic activities [1–3]. Moreover, the use of essential oils as natural antioxidants attracts the interest of many scientists and researchers due to their utility in the prophylaxis and treatment of diseases related to oxidative stress [4].

Analytically speaking, an essential oil is usually regarded as a complex sample comprising a wide variety of volatile compounds strictly embedded with each other in a matrix. To unravel such an intricate composition, the high separation power of gas chromatography works as the most effective technique. After the widespread diffusion of essential oils in numerous different fields of current society, a detailed legislation on their global marketing has been issued. As a consequence, the need for reliable quantitative data has grown even more due to quality and safety requirements concerns. As well, an interest in essential oils' biological activity is continually growing [5]. The need for accurate quantitation in essential oils analysis is not only dictated by regulatory bodies (IFRA, EFSA, etc.) but becomes fundamental in novel characterization. Although relative percentage abundance is the most widespread approach to quantification by GC, this method is often improperly applied, e.g., for comparison of different essential oil samples from the same species [6]. In order to be compared, data first need to be standardized. The employment of calibration procedures with standards is even more important when true quantitation in physical units of concentration is the purpose of analysis. In the present study, response factors measured by means of external and internal methods were applied to true quantitation of the volatiles released by *Senecio anteuphorbium*. The genus *Senecio* comprises almost 1100 species distributed in Asia and Africa. In Morocco, *Senecio anteuphorbium* (syn. *Kleinia anteuphorbium* (L.) DC.), an endemic medicinal plant locally called "Achbartou", is commonly used in traditional medicine, mainly in the Sousse region, as a sedative for abdominal or back pain [7]. In fact, this fatty plant is also used as an anti-inflammatory, hemostatic, and in the treatment of rheumatism [8,9]. However, this plant species has received very little attention as regards its composition and biological activity. Beyond the chemistry, the scope of the study was to explore the allelopathic properties of the essential oil against *Lactuca sativa* germination and growth. In fact, a number of reports have previously highlighted the phytotoxic activity of the genus *Senecio* [10–15].

Allelopathy, a phenomenon that is gaining much interest in plant science, works via a complex mechanism. Plants contain several organic molecules that are biologically active. These molecules are associated with active roles in plant defense against pests, herbivores, and pathogens, as well as environmental stresses [16]. The production and concentration of secondary metabolites in plants are multifaceted processes that depend on the plant species, age, organs, and biotic and abiotic stresses they are subjected to [17]. Allelochemicals are liberated by plants into the outer environment and interact with close plants as well as other organisms corresponding to the suppression or stimulation of the growth, physiology, and development of the target species [18]. Briefly, allelopathy falls within the more recent and sustainable strategies applied to weed suppression in substitution of synthetic organic herbicides.

To the best of our knowledge, this work represents the first comprehensive investigation into the volatile constituents of *S. anteuphorbium*.

2. Materials and Methods

2.1. Plant Material

The aerial parts of *S. anteuphorbium* were harvested during April 2019 (flowering stage) from Taghazout region (30°31′59″ N, 9°42′00″ W). The collected plant was air-dried in the shade for ten days and ground into a fine powder. A voucher specimen (No. 10119) was deposited in the Herbarium of Laboratory of Agri-Food, Biotechnologies and Valorization of Plant Bioresources, Faculty of Science, Marrakesh.

2.2. Extraction of the Essential Oil

The EO of the air-dried aerial parts of *S. anteuphorbium* (3 × 300 g) was extracted by hydro-distillation using a Clevenger-type apparatus (ENVEA, Casablanca, Morocco) for four hours by setting an initial temperature of 100 °C, decreased to a constant temperature of 70 °C after 30 min. The essential oil obtained was dried with sodium sulfate (Sigma-

Aldrich, St Louis, MO, USA) and stored at 4 °C until further analysis. The oil yield was calculated according to the following formula:

$$Y\% = \frac{V_{EO}}{DW} \times 100$$

where V_{EO}: volume of essential oil recovered (mL); and DW: amount of dry plant material used for extraction (g).

2.3. Gas Chromatography

2.3.1. GC–FID Analysis

Senecio anteuphorbium essential oils were injected into a Shimadzu GC-2010 system equipped with an AOC-20i autosampler and a split/splitless injector. The analytical column was a Zebron-5 ms, 30 m × 0.25 mm i.d. × 0.25 μm film thickness (Phenomenex, Los Angeles, CA, USA). The oven program was: 50 °C, held for 1 min, at 3 °C/min to 250 °C, held 5 min. Injection temperature and volume were 250 °C and 1.0 μL, respectively. Samples were previously diluted 1:10 v/v in n-hexane. Injection mode: split, with a split ratio 1:50. Carrier gas was helium (u, 30 cm·s^{-1}; inlet pressure, 99.0 kPa). Detector temperature: 300 °C. Detector gases: H_2, 40 mL·min^{-1}; air, 400 mL·min^{-1}. Data were handled by means of *GCsolution* software (Shimadzu, Japan). For the calculation of response factors, the volatiles identified by GC–MS were grouped according to their structure (hydrocarbon, alcohol, ketone, etc.), and, for each group, a representative chemical was chosen and calibrated according to the following procedure. The reference standard (e.g., β-bisabolene for sesquiterpene hydrocarbons) was externally and internally calibrated by injecting 5 different levels of concentration within the linear range with the addition of a fixed amount of nonane as internal standard (final concentration 0.1 g/100 g). Each level of concentration was analyzed in triplicate (Table 1). The response factor was measured according to:

$$\boldsymbol{RF} = \frac{[\mathbf{Std}]\,[\boldsymbol{i.s.}]}{\mathbf{Area\ std} : \boldsymbol{Area\ i.s.}}$$

where ***RF***, response factor; **[Std]**, concentration (g/100 g) of the standard to be calibrated; **[*i.s.*]**, concentration of the internal standard (g/100 g); **Area std**, FID peak area of the standard; ***Area i.s.***, FID peak area of the internal standard (n-nonane). The concentration of each volatile component in essential oils real samples was then calculated:

$$[\boldsymbol{VOC}] = \frac{(\boldsymbol{Area}_{\mathrm{VOC}} : \boldsymbol{Area\ i.s.})\cdot\boldsymbol{RF}\cdot[\boldsymbol{i.s.}]\cdot 100}{\boldsymbol{W}_{oil}}$$

where $\boldsymbol{Area}_{\mathrm{VOC}}$, FID peak area of each volatile constituent; $\boldsymbol{W}_{oil}$, weight (g) of the oil.

Table 1. Composition of *Senecio anteuphorbium* essential oil. Values are means of triplicate analyses.

Peak#	Group	Std.	Compound	RI_{exp}	RI_{db}	Area %	RF	Wt% (g/100 g)
1	H	√	α-Pinene	933	933	0.16 ± 0.01	1.0	0.15 ± 0.01
2	A	√	1-Octen-3-ol	978	978	0.32 ± 0.03	1.3	0.38 ± 0.03
3	A	√	Linalool	1099	1101	0.26 ± 0.01	1.3	0.31 ± 0.01
4	K	√	Isophorone	1122	1123	0.12 ± 0.01	1.3	0.14 ± 0.01
5	A		trans-Sabinol	1139	1140	0.32 ± 0.01	1.3	0.38 ± 0.01
6	A		trans-Verbenol	1142	1145	0.47 ± 0.02	1.3	0.56 ± 0.02
7	O	√	trans-Linalool oxide	1173	1174	0.49 ± 0.04	1.5	0.67 ± 0.05
8	A	√	p-Cymen-8-ol	1187	1189	0.62 ± 0.01	1.3	0.73 ± 0.01
9	A	√	Myrtenol	1199	1202	0.37 ± 0.02	1.3	0.44 ± 0.02

Table 1. *Cont.*

Peak#	Group	Std.	Compound	RI_{exp}	RI_{db}	Area %	RF	Wt% (g/100 g)
10	K	√	Verbenone	1206	1208	0.28 ± 0.01	1.3	0.33 ± 0.01
11	AL	√	Safranal	1208	1201	0.28 ± 0.01	1.4	0.36 ± 0.01
12	K	√	Pulegone	1242	1241	0.10 ± 0.02	1.3	0.12 ± 0.02
13	K	√	Carvenone	1255	1257	0.29 ± 0.02	1.3	0.34 ± 0.02
14	A	√	4-Vinylguaiacol	1308	1309	0.77 ± 0.02	1.3	0.91 ± 0.02
15	H		Silphiperfol-5-ene	1329	1326	0.70 ± 0.02	1.0	0.64 ± 0.02
16	H	√	δ-Elemene	1337	1335	0.28 ± 0.01	1.0	0.25 ± 0.01
17	H		Presilphiperfol-7-ene	1342	1339	0.39 ± 0.01	1.0	0.35 ± 0.01
18	A	√	trans-p-Menth-6-en-2,8-diol	1375	1375	1.55 ± 0.04	1.3	1.83 ± 0.05
19	H		Silphiperfol-6-ene	1382	1380	0.22 ± 0.02	1.0	0.20 ± 0.02
20	H	√	β-Patchoulene	1385	1383	0.66 ± 0.03	1.0	0.60 ± 0.02
21	H	√	α-Copaene	1386	1385	1.72 ± 0.03	1.0	1.56 ± 0.03
22	H		Modhephene	1388	1384	0.71 ± 0.02	1.0	0.65 ± 0.01
23	H	√	β-Elemene	1391	1389	0.71 ± 0.01	1.0	0.65 ± 0.01
24	H		α-Isocomene	1393	1387	0.36 ± 0.02	1.0	0.33 ± 0.02
25	H		α-Gurjunene	1409	1406	0.34 ± 0.01	1.0	0.31 ± 0.01
26	H		β-Isocomene	1411	1407	0.29 ± 0.01	1.0	0.26 ± 0.01
27	H	√	(Z)-Caryophyllene	1415	1413	0.37 ± 0.02	1.0	0.34 ± 0.02
28	H	√	Guaia-6,9-diene	1445	1444	0.21 ± 0.02	1.0	0.19 ± 0.01
29	H	√	α-Humulene	1456	1454	0.41 ± 0.02	1.0	0.37 ± 0.02
30	H	√	Alloaromadendrene	1460	1458	0.51 ± 0.02	1.0	0.46 ± 0.02
31	H		Selina-4,11-diene	1478	1476	5.59 ± 0.13	1.0	5.08 ± 0.11
32	H	√	Germacrene D	1482	1480	1.87 ± 0.02	1.0	1.70 ± 0.02
33	H		Aristolochene	1491	1487	0.55 ± 0.02	1.0	0.50 ± 0.01
34	H	√	Bicyclogermacrene	1501	1497	25.02 ± 0.09	1.0	22.75 ± 0.08
35	A		Cubebol	1520	1519	0.21 ± 0.02	1.3	0.25 ± 0.02
36	H	√	δ-Cadinene	1523	1518	0.39 ± 0.03	1.0	0.35 ± 0.03
37	A		α-Elemol	1552	1546	0.12 ± 0.01	1.3	0.14 ± 0.01
38	A	√	Spathulenol	1577	1576	21.37 ± 0.36	1.3	25.26 ± 0.42
39	A		Fokienol	1596	1596	4.96 ± 0.11	1.3	5.86 ± 0.13
40	A		1,10-di-epi-Cubenol	1616	1614	1.32 ± 0.10	1.3	1.56 ± 0.12
41	A		epi-γ-Eudesmol	1627	1624	5.75 ± 0.17	1.3	6.80 ± 0.20
42	A		T-muurolol	1652	1645	0.35 ± 0.04	1.3	0.41 ± 0.05
43	A		Cadin-4-en-10-ol	1661	1659	1.04 ± 0.04	1.3	1.23 ± 0.04
44	A		Shyobunol	1690	1686	1.74 ± 0.07	1.3	2.06 ± 0.08
45	A		β-Acoradienol	1763	1760	0.25 ± 0.02	1.3	0.30 ± 0.02
46	K		Aristolone	1765	1759	0.81 ± 0.02	1.3	0.96 ± 0.02
			HYDROCARBONS			41.46 ± 0.31		37.69 ± 0.38
			ALCOHOLS			41.79 ± 0.47		49.38 ± 0.37
			KETONES			1.60 ± 0.07		1.89 ± 0.08
			ALDEHYDES			0.28 ± 0.02		0.36 ± 0.03
			OXIDES			0.49 ± 0.02		0.67 ± 0.04

Group: H, hydrocarbon; A, alcohol; K, ketone; AL, aldehyde; O, oxide. Std.: co-injection of reference standard; RI_{exp}: retention index experimentally determined against a mixture of n-alkanes (C7-C40) on a Zebron-5 ms column. RI_{db}: retention index retrieved from FFNSC 2 and Adams 4th edition databases. RF: response factor.

The reference standards β-caryophyllene, caryophyllene oxide, farnesol, and citronellal, all supplied by Merck, were chosen as representative compounds of the different chemical groups.

2.3.2. GC–MS Analysis

Qualitative analyses were carried out on a GCMS-TQ8030 (Shimadzu) equipped with a Zebron-5ms column (30 m × 0.25 mm i.d. × 0.25 μm film thickness). Carrier gas (He) parameters were the same as for GC–FID analysis. MS conditions: interface and source temperatures, 230 °C and 200 °C, respectively. Ionization mode: EI, 0.9 kV; acquisition mass range, 40–400 m/z; scan speed, 1666 amu/s; scan interval, 0.25 s. Data were handled by means of GCMSsolution software. Identification was carried out by matching unknown spectra with two databases (FFNSC2, Adams 4th edn.). Moreover, retention indices were calculated by injecting a mix of saturated alkanes ranging from heptane to tetracontane (Merck, Darmstadt, Germany). Experimental retention indices were then compared with those listed in databases in order to restrict the list of candidates.

2.4. Allelopathic Activity

The allelopathic activity of *S. anteuphorbium* EO was evaluated against lettuce seeds according to Jalaei et al., with slight modifications [19]. In brief, commercial lettuce (*Lactuca sativa*) seeds were surface sterilized with 70% ethanol for 30 s, washed with sterile water to remove the ethanol, and then disinfected for 20 min by a 0.2% sodium hypochlorite solution, followed by three rinses of distilled water for 5 min each. After disinfection, 30 seeds were placed in glass Petri dishes lined with a filter paper Whatman No.1. The seeds were soaked with 10 mL of each EO concentration prepared in dimethyl sulfoxide (DMSO) (1%, v/v). The Petri dishes were sealed with parafilm and incubated in a climate room at 25 ± 1 °C with photoperiod 12:12 for 10 days [19]. The experience was performed in quadruplet, and a control containing DMSO 1% was added. Observations and measurements were made on the germination rate of seeds and the growth of lettuce seedlings. Plant growth in the incubation period was expressed by root length (cm), shoot length (cm), and seedling length (cm) at the end of the period. The shoot and root lengths of all seedlings per each plate were measured, and the allelopathic inhibition of root and shoot growth was calculated, with respect to control, as follows:

$$Inhibition\ \% = 100 * (No/Length\ of\ control - No/Length\ of\ treatment)/No/Length\ of\ control$$

The germination and seedling growth were also evaluated by percentage of germination according to the following formula:

$$Germination\ percentage = (n/N) * 100$$

with n: number of germinated seeds in each concentration, and N: total number of seeds (30).

Moreover, mean germination time (MGT), germination rate (GR), and the vigor index (VI) were calculated.

3. Results and Discussion

3.1. Quantitative Analysis

The hydro-distillation extraction of the aerial parts of *S. anteuphorbium* from Taghazout region yielded 0.3% (v/w) of oil distillates. This EO was characterized by a dark brown color, an oily appearance, a density of 0.9 g/mL, and a freezing point above −21 °C. With regard to the yield, 0.15% was that obtained from *S. anteuphorbium* harvested in Essaouira region [20]. The higher yield of *S. anteuphorbium* from Taghazout (actual samples) can be explained by the different geographic provenance (different pedoclimatic conditions).

Figure 1 shows the GC–MS fingerprint of *S. anteuphorbium* EO, whereas its composition is reported in Table 1. The precision of the GC methodology was tested through the measurement of standard deviations (reported in Table 1) and coefficients of variation, always lower than 5%. As can be seen, based on peak area normalization, the total identified and quantified fraction amounts to 85.62% of the whole EO. The volatiles were equally distributed between terpene hydrocarbons (41.46%) and alcohols (41.79%), with minor constituents being ketones, aldehydes, and oxides. *S. anteuphorbium* can be safely described as a sesquiterpene-rich oil, reporting as predominant components bicyclogermacrene (25.02%) and spathulenol (21.37%), a sesquiterpene alcohol. To confirm this, the chromatogram results are much more crowded in the second region, which is a typical elution zone of sesquiterpenes in essential oil analysis. Worth mentioning is the presence of characteristic components, such as silphiperfolene isomers, α- and β-isocomene, aristolochene, aristolone, and shyobunol. This chemical description matches roughly with that published by Elhidar et al. [20], the sole report on *S. anteuphorbium* found in the literature and totally focused on this species. However, the authors carried out only a tentative identification of the components by mass spectral matching with wide and generic collections (Wiley and NIST); in the present study, the mass spectral libraries used (FFNSC and Adams) were exclusively pertinent to flavor and fragrance compounds. Additionally, the identification procedure was boosted by the retention index matching [21] tool that was only mentioned but not implemented in Elhidar et al. [20]. Finally, an accurate calibration in absolute units (weight%, g/100 g) of single components was here carried out, enhancing the value of quantitative analysis [22]. There is evidence in the literature of a considerable number of papers on *Senecio* spp. Nonetheless, only a few publications are based on accuracy and reliability of both the data produced and methodologies used since, in many of them, the determination of artifacts (not naturally occurring compounds) is widely manifest [14,23,24]. For example, Irahal et al. have recently reported a GC–MS analysis of a variety of plant species, including *S. anteuphorbium* [24]. The GC data presented in that study show a lack of reliability in the GC–MS analytical protocol employed. Some compounds were artifacts (i.e., cyclotrisiloxane); many others were simply misidentified (i.e., β-maaliene, which is a sesquiterpene, notoriously eluting after monoterpenes); the mass spectral library used was only a NIST, capable of assigning an identity to a lower fraction of peaks (79%); finally, the source of retention indices was not specified. Nonetheless, a comparison between the two volatile fingerprints has been made, highlighting a substantial difference as regards the predominant constituents. Less than ten components were commonly found in the two compositions, among which selina-4,11-diene and shyobunol are worthy to be mentioned. Moreover, in many cases, these reports have been published in journals that do not enjoy scientific prestige. To further discuss the chemical constituents of the present *S. anteuphorbium* samples, it seems worthwhile to emphasize the similarity found with other *Senecio* spp. Spathulenol and germacrene B, which dominated the volatile fingerprint in *S. rowleyanus* and also showed powerful antioxidant and antimicrobial activities [25].

Rich in sesquiterpenes (i.e., germacrene D and A, γ- and δ-cadinene) were the essential oils from the roots and leaves of *S. rufinervis* [26]. The composition of *S. vernalis* was the most similar to that of *S. anteuphorbium*, bicyclogermacrene and spathulenol being the major volatiles, as here [27]. *S. trapezuntinus* appeared as a sesquiterpene rich oil, with (E)-β-farnesene as the predominant volatile, but the variety of minor sesquiterpenes was comparable to that reported by *S. anteuphorbium* [28]. Leaving aside the specific composition of *S. anteuphorbium*, it should be pointed out that numerous papers establish a close correlation between the sesquiterpene component and allelopathy [29,30]. Allelopathy, the natural phenomenon of plants interaction, represents great promise in crop science, with most allelochemicals being valid herbicides/pesticides in support of the sustainable development of agriculture. Among allelochemicals, spathulenol and bicyclogermacrene have been previously determined in a variety of plant species [31,32], confirming the present findings with regard to the allelopathic assays.

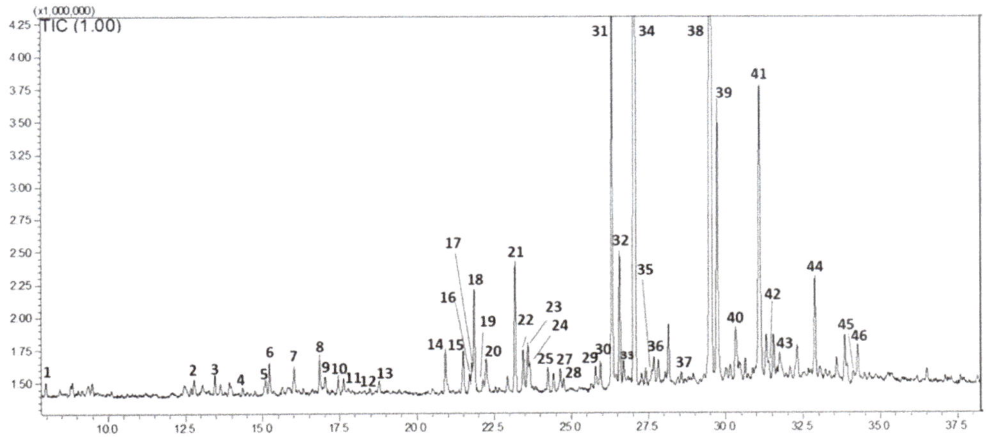

Figure 1. Total ion chromatogram obtained from GC–MS analysis of *S. anteuphorbium* essential oil. Peak top numbers refer to compounds listed in Table 1.

3.2. Allelopathic Activity

The allelopathic potential of the *S. anteuphorbium* EO was evaluated through germination percentage, root, shoot and seed lengths, and inhibitory growths percentage and factors on *Lactuca sativa* seeds. The results showed (Table 2, Figures 2 and 3) that the extracted EO from *S. anteuphorbium* exhibited a significant allelopathic inhibitory effect on the germination and the seedling growth of the tested seeds (*Lactuca sativa*) in a dose-dependent manner compared to the control.

Table 2. Effects of different concentrations of *S. anteuphorbium* essential oil on shoot, root, and seed germination indices and seedling growth factors of L. sativa at the end of incubation time. GP: germination percentage, MGT: mean germination time, GR: germination rate, VI: vigor index.

EO Concentration (mg/mL)	Shoot Growth (cm)	Root Growth (cm)	Seedling Growth (cm)	GP (%)	MGT (Day)	GR (Units)	VI (Units)
Control	2.34 ± 0.072 ^{a}	3.41 ± 0.121 ^{a}	5.61 ± 0.1 ^{a}	100 ^{a}	1.93 ± 0.14 ^{a}	8.91 ± 1.47 ^{a}	560.64 ± 12.47 ^{a}
0.018	2.33 ± 0.12 ^{a}	3.28 ± 0.21 ^{a}	5.76 ± 0.33 ^{a}	45.55 ± 6.94 ^{b}	2.01 ± 0.33 ^{b}	6.01 ± 0.57 ^{b}	260.95 ± 12.23 ^{b}
0.035	1.85 ± 0.12 ^{b}	2.7 ± 0.16 ^{a}	4.55 ± 0.21 ^{b}	15.55 ± 1.92 ^{c}	3.01 ± 0.52 ^{c}	1.7 ± 0.02 ^{c}	70.81 ± 7.17 ^{c}
0.07	1.06 ± 0.1 ^{b}	2.23 ± 0.21 ^{b}	3.29 ± 0.16 ^{b}	8.88 ± 1.92 ^{cd}	3 ± 0.64 ^{c}	1.05 ± 0.28 ^{d}	27.01 ± 13.37 ^{cd}
0.14	0.55 ± 0.02 ^{c}	1.28 ± 0.04 ^{b}	1.83 ± 0.03 ^{c}	7.77 ± 1.92 ^{de}	3.11 ± 0.51 ^{c}	0.92 ± 0.15 ^{d}	14.23 ± 1.89 ^{d}
0.28	0.29 ± 0.02 ^{d}	0.72 ± 0.01 ^{c}	1.01 ± 0.02 ^{c}	3.33 ^{e}	3.33 ± 0.8 ^{c}	0.4 ± 0.1 ^{e}	3.36 ± 0.76 ^{e}
IC_{50}	0.12 ± 0.01	0.15 ± 0.02	-	-	-	-	-

IC_{50} is the concentration at which 50% of growth inhibition was predicted to occur. Different letters indicate statistically significant differences at $p \leq 0.05$. Significance level: ^{a} $p > 0.05$, ^{b} $p < 0.05$, ^{c} $p < 0.01$, ^{d} $p < 0.001$, and ^{e} $p < 0.0001$.

According to the obtained results reported in Table 2 and Figure 3, at a high concentration of the EO (0.28 mg/mL), the growth of shoots and roots was reduced by 87.60% and 78.88%, respectively. The EO showed an IC_{50} value of 0.12 mg/mL for shoot growth and 0.15 µg/mL for root growth, respectively, compared to the controls. As regards mean germination time (MGT), although the results show a statistically significant difference between the control and treated samples, MGT was the least affected parameter. The germination rate (GR) differed significantly between the different treatments, the highest being observed with the control (8.91 units) and the lowest being observed with *S. an-*

teuphorbium EO at 0.28 mg/mL with 0.4 units. Consequently, the GR factor was strictly concentration-dependent, as well as the vigor index (3.36 vs. 560.64, for 0.28 mg/mL EO and the control, respectively).

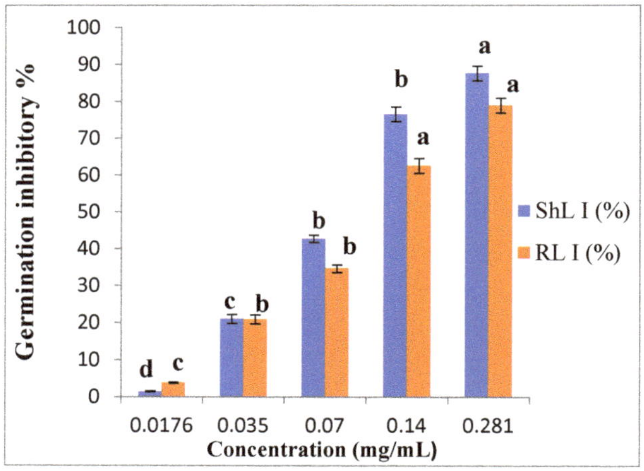

Figure 2. Allelopathic effect of the essential oil from *Senecio anteuphorbium* on root (RLI) and shoot (ShLI) growths of *Lactuca sativa* seeds. Different letters indicate statistically significant differences at $p \leq 0.05$.

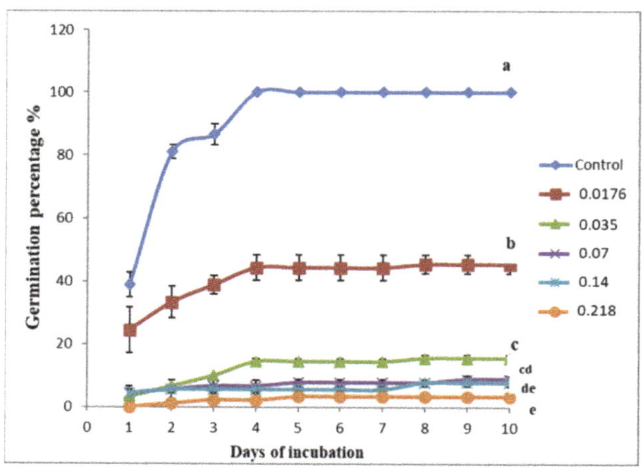

Figure 3. Cumulative germination observed in *L. sativa* seeds after treatment with different EO concentrations (mg/mL) of *Senecio anteuphorbium* at 1-day intervals.

The comprehensive observation of germination data after treatments at 1-day intervals (Figure 3) shows that all the EO concentrations inhibited germination compared with the control. The maximum germination percentage (100%) was recorded in the control and the minimum (3.33%) in the EO treatment (0.28 mg/mL) at the end of the incubation period. The genus *Senecio* has been widely reported in the literature due to the phytocompounds found in extracts, with a potent allelopathic property. The phytotoxic effects of *S. westermanii* Dusén ethanol extracts and subfractions were evaluated on lettuce (*Lactuca sativa* L.) and onion (*Allium cepa* L.) seeds. The results demonstrated an inhibitory effect on the germination velocity index and growth. The plant also caused a change in respiration and photosynthesis [13]. Moreover, it was observed by Arancibia and his coauthors [14] that aqueous and ethanolic extracts, from *S. filaginoides*, significantly inhibited the germination of *Solanum lycopersicum* and *Lolium multiflorum* seeds in a dose-dependent manner. Cruz-Silva and his collaborators demonstrated that aqueous extracts of leaves and inflorescences of *S. brasiliensis* altered the germination and seedling development of *Lactuca sativa* L. (lettuce) [11]. The phytotoxic effectiveness of the *Senecio* genus has been poorly studied. To the best of our knowledge, the results obtained in this study are the first published data concerning the allelopathic activity of *Senecio anteuphorbium* essential oil. Compared to other species of the genus *Senecio*, the phytotoxic effects obtained in our study are higher than the allelopathic potential of EOs from the *Senecio* genus previously evaluated in other studies. Overall, it has been previously shown that the essential oil of *S. amplexicaulis* demonstrated remarkable and dose-dependent phytotoxic activity at the tested concentration and significant reduction in seed germination percentage of *Phalaris minor* and *Triticum aestivum* at 0.5 mg/mL; the oil inhibited germination of both *T. aestivum* (65.00 $\pm$ 1.00%) and *P. minor* (58.34 $\pm$ 1.52%) compared to the control [12]. In another study, the phytotoxic activity of the essential oil of *Senecio erucifolius* L. was evaluated against three weeds, namely *Medicago sativa* L., *Urtica cannabina* L., and *Amaranthus retroflexus* L. [15]. In general, the essential oil at low concentrations stimulated the growth of sprouts of the treated species; at high concentrations, it suppressed it. The length of the roots of *M. sativa*, *U. cannabina*, and *A. retroflexus* increased by 21.00, 10.46, and 2.53%, respectively, after treatment with oil at the lowest concentration (0.125 mg/mL) and decreased by 9.36, 23.00, and 19.53% after treatment at the highest concentration (4 mg/mL) [15]. In fact, the change in the parameters of germination indicates the changes in the metabolic reactions of the plants [33]. It might be speculated that the phytotoxic effect of *S. anteuphorbium* EO is due to the activity of the major compounds, bicyclogermacrene (25.02%) and spathulenol (21.37%) (Table 1). However, minor compounds, such as α-pinene and p-cymen-8-ol, have been reported as allelochemicals, and they could act either individually or synergistically as inhibitors on the germination and growth of the *Lactuca sativa* L. seeds [34]. Moreover, the oxygenated terpenoids usually have a significant role in biological activity compared to non-oxygenated compounds due to oxygen reactivity [29].

4. Conclusions

This article provides a dataset on the volatile quantitative composition and allelopathic properties of *S. anteuphorbium*, an endemic plant of Morocco used in traditional medicine. The EO composition was thoroughly investigated by GC techniques, applying true quantitation in order to overcome the limits of traditional approaches based on the sole relative abundance. The EO reported a predominant sesquiterpene component. A strong allelopathic effect against *Lactuca sativa* seeds was observed, suggesting an interesting application of the EO as an alternative bioherbicide. All the findings suggest the usefulness of further studies (i.e., investigation of the non-volatile fraction, antioxidant power, and mode of action) on this *Senecio* species.

Author Contributions: Conceptualization, M.M., K.B. and M.L.; methodology, S.O. and A.A.; software, R.C. and R.V.; validation, M.M. and M.L.; formal analysis, S.O., W.B.B., A.A. and S.A.; investigation, S.O. and S.S.; resources, S.A. and R.V.; data curation, S.O., S.S., S.A., R.C. and R.V.; writing—original draft preparation, S.O. and W.B.B.; writing—review and editing, R.C., S.O., W.B.B., A.A. and S.A.; supervision, K.B., M.L., N.C. and M.M. All authors have read and agreed to the published version of the manuscript.

Funding: This research received no external funding.

Acknowledgments: This work was supported by the "Centre National de Recherche Scientifique et technique", Morocco, (URAC 35).

Conflicts of Interest: The authors declare no conflict of interest.

References

1. Brahim, M.A.S.; Fadli, M.; Hassani, L.; Boulay, B.; Markouk, M.; Bekkouche, K.; Abbda, A.; Ait Ali, M.; Larhsini, M. *Chenopodium ambrosioides* var. ambrosioides used in Moroccan traditional medicine can enhance the antimicrobial activity of conventional antibiotics. *Ind. Crop. Prod.* **2015**, *71*, 37–43. [CrossRef]
2. Do Socorro Costa, M.; Santos Araújo, N.J.; Sampaio de Freitas, T.; Bezerra da Cunha, F.A.; do Amaral, W.; Deschamps, C.; Confortin, C.; Everson da Silva, L.; Douglas Melo Coutinho, H. GC–FID Analysis and Antibacterial Activity of the *Calyptranthes concinna* Essential Oil against MDR Bacterial Strains. *Separations* **2020**, *7*, 10. [CrossRef]
3. Aghraz, A.; Albergamo, A.; Benameur, Q.; Salvo, A.; Larhsini, M.; Markouk, M.; Gervasi, T.; Cicero, N. Polyphenols contents, heavy metals analysis and in vitro antibacterial activity of extracts from *Cladanthus arabicus* and *Bubonium imbricatum* of Moroccan Origin. *Nat. Prod. Res.* **2020**, *34*, 63–70. [CrossRef] [PubMed]
4. Bakkali, F.; Averbeck, S.; Averbeck, D.; Idaomar, M. Biological effects of essential oils—A review. *Food Chem. Toxicol.* **2008**, *46*, 446–475. [CrossRef] [PubMed]
5. Mladenova, T.; Stoyanov, P.; Todorov, K.; Davcheva, D.; Kirova, G.; Deneva, T.; Gyuzeleva, D.; Mladenov, R.; Bivolarska, A. Phytochemical and Biological Traits of Endemic *Betonica bulgarica* (Lamiaceae). *Separations* **2021**, *8*, 11. [CrossRef]
6. Bicchi, C.; Liberto, E.; Matteodo, M.; Sgorbini, B.; Mondello, L.; d'Acampora Zellner, B.; Costa, R.; Rubiolo, P. Quantitative analysis of essential oils: A complex task. *Flavour Fragr. J.* **2008**, *23*, 382–391. [CrossRef]
7. Cicuzza, D.; Stäheli, D.S.; Nyffeler, R.; Eggli, U. Morphology and Anatomy Support a Reclassification of the African Succulent Taxa of *Senecio* S.L. (Asteraceae: Senecioneae). *Haseltonia* **2017**, *23*, 11–26. [CrossRef]
8. Bellakhdar, J. La pharmacopée marocaine traditionnelle. In *Médecine Arabe Ancienne et Savoirs Populaires;* Ibis Press: Paris, France, 1997; pp. 1–764.
9. Lounis, H.; Bergheim, I.; Bouhaimi, A.; Guigonis, J.M.; Belhamel, K. Anti-Inflammatory and antioxidant activities of *Inula viscosa* and *Senecio anteuphorbium*. *Orient. Pharm. Exp. Med.* **2018**, *18*, 225–236. [CrossRef]
10. Ahmed, M.; Wardle, D.A. Allelopathic potential of vegetative and flowering ragwort (*Senecio jacobaea* L.) plants against associated pasture species. *Plant Soil* **1994**, *164*, 61–68. [CrossRef]
11. Cruz-Silva, C.T.A.; Santorum, M.; Bini, F.V. Efeito alelopatico de extratos aquosos de *Senecio brasiliensis* (Spreng) Less sobre a germinação e o desenvolvimento de plântulas. *Cultiv. Saber* **2009**, *2*, 62–70.
12. Singh, R.; Ahluwalia, V.; Singh, P.; Kumar, N.; Prakash Sati, O.; Sati, N. Antifungal and phytotoxic activity of essential oil from root of *Senecio amplexicaulis* Kunth. (Asteraceae) growing wild in high altitude-Himalayan region. *Nat. Prod. Res.* **2016**, *30*, 1875–1879. [CrossRef] [PubMed]
13. Merino, F.J.Z.; Ribas, D.F.; Silva, C.B.D.; Duarte, A.F.S.; Paula, C.D.S.; Oliveira, M.D.; Miguel, O.G. A study of the phytotoxic effects of the aerial parts of *Senecio westermanii* Dusén (Asteraceae) on *Lactuca sativa* L. and *Allium cepa* L. seeds. *Brazil. J. Pharm. Sci.* **2018**, *54*, e17135. [CrossRef]
14. Arancibia, L.A.; Henriquez, A.M.; Marchiaro, A.B. *Senecio filaginoides* DC as a Source of Allelopathic Agents and Its Possible Use as a Bioherbicide. *Asian Res. J. Curr. Sci.* **2020**, *2*, 101–107.
15. Zhang, C.P.; Toshmatov, Z.O.; Zhou, S.X.; Li, W.J.; Zhang, C.; Shao, H. Chemical Composition and Phytotoxic Activity of Essential Oil from *Senecio erucifolius*. *Chem. Nat. Compd.* **2021**, *57*, 580–582. [CrossRef]
16. Akula, R.; Ravishankar, G.A. Influence of abiotic stress signals on secondary metabolites in plants. *Plant Signal. Behav.* **2011**, *6*, 1720–1731. [CrossRef]
17. Siyar, S.; Majeed, A.; Muhammad, Z.; Ali, H.; Inayat, N. Allelopathic effect of aqueous extracts of three weed species on the growth and leaf chlorophyll content of bread wheat. *Acta Ecol. Sin.* **2019**, *39*, 63–68. [CrossRef]
18. Gniazdowska, A.; Bogatek, R. Allelopathic interactions between plants. Multi-site action of allelochemicals. *Acta Physiol. Plant.* **2005**, *27*, 395–407. [CrossRef]
19. Jalaei, Z.; Fattahi, M.; Aramideh, S. Allelopathic and insecticidal activities of essential oil of *Dracocephalum kotschyi* Boiss. from Iran: A new chemotype with highest limonene-10-al and limonene. *Ind. Crop. Prod.* **2015**, *73*, 109–117. [CrossRef]
20. Elhidar, N.; Nafis, A.; Kasrati, A.; Goehler, A.; Bohnert, J.A.; Abbad, A.; Mezrioui, N.E. Chemical composition, antimicrobial activities and synergistic effects of essential oil from *Senecio anteuphorbium*, a Moroccan endemic plant. *Ind. Crop. Prod.* **2019**, *130*, 310–315. [CrossRef]
21. Costa, R.; De Fina, M.R.; Valentino, M.R.; Dugo, P.; Mondello, L. Reliable identification of terpenoids and related compounds by using Linear Retention Indices interactively with Mass Spectrometry search. *Nat. Prod. Commun.* **2007**, *2*, 413–418. [CrossRef]
22. Costa, R.; Zellner, B.; Crupi, M.L.; De Fina, M.R.; Valentino, M.R.; Dugo, P.; Dugo, G.; Mondello, L. Gas Chromatography-Mass Spectrometry (GC–MS), Gas Chromatography-Olfactometry (GC-O) and Enantio-GC Investigation on the Essential Oil of *Tarchonanthus camphoratus* L. *Flavour Fragr. J.* **2008**, *23*, 40–48. [CrossRef]
23. Kebbi, S.; Noman, L.; Demirtas, I.; Bensonici, C.; Adem, S.; Benayache, S.; Benayache, F.; Seghiri, R.; Gok, M. In Vitro Antioxidant and Anticholinesterase Activities of *Senecio massaicus* Essential Oil and Its Molecular Docking Studies as a Potential Inhibitor of COVID-19 and Alzheimer's Diseases. *J. Biolog. Act. Prod. Nat.* **2021**, *11*, 380–394. [CrossRef]
24. Irahal, I.N.; Lahlou, F.A.; Hmimid, F.; Errami, A.; Guenaou, I.; Diawara, I.; Kettani-Halabi, M.; Fahde, S.; Ouafik, L.; Bourhim, N. Identification of the chemical composition of six essential oils with mass spectroscopy and evaluation of their antibacterial and antioxidant potential. *Flavour Fragr. J.* **2021**, *36*, 465–476. [CrossRef]

25. ElHawary, S.S.; Galal, A.E.; Yousif, M.F.; Kirollos, F.N. GC–MS and bioactivity of the essential oil of *S. rowleyanus* Jacobs. *Pharmacog. Magaz.* **2008**, *4*, 273–277.
26. Thakuri, B.C.; Padalia, R.C.; Chanotiya, C.S.; Tiwari, A.; Mathela, C.S. Sesquiterpene rich oils from leaves and roots of *S. rufinervis* DC. *J. Essent. Oil Res.* **2008**, *20*, 435–436. [CrossRef]
27. Nori-Shargh, D.; Raftari, S.; Deyhimi, F. Analysis of the essential oil of *Senecio vernalis* Waldst. & Kit. from Iran. *Flavour Fragr. J.* **2008**, *23*, 357–359. [CrossRef]
28. Üçüncü, O.; Yayli, N.; Yaşar, A.; Terzioğlu, S.; Yayli, N. Chemical composition of the essential oils from flower, leaf, and stem of *S. trapezuntinus* Boiss. grown in Turkey. *Nat. Prod. Commun.* **2008**, *3*, 925–928. [CrossRef]
29. Assaeed, A.; Elshamy, A.; El Gendy, A.E.N.; Dar, B.; Al-Rowaily, S.; Abd-ElGawad, A. Sesquiterpenes-Rich essential oil from above ground parts of *Pulicaria somalensis* exhibited antioxidant activity and allelopathic effect on weeds. *Agronomy* **2020**, *10*, 399. [CrossRef]
30. Landi, M.; Misra, B.B.; Muto, A.; Bruno, L.; Araniti, F. Phytotoxicity, morphological, and metabolic effects of the sesquiterpenoid nerolidol on *Arabidopsis thaliana* seedling roots. *Plants* **2020**, *9*, 1347. [CrossRef]
31. Hamrouni, L.; Hanana, M.; Amri, I.; Romane, A.E.; Gargouri, S.; Jamoussi, B. Allelopathic effects of essential oils of *Pinus halepensis* Miller: Chemical composition and study of their antifungal and herbicidal activities. *Arch. Phytopathol. Plant Prot.* **2015**, *48*, 145–158. [CrossRef]
32. Candido, L.P.; Varela, R.M.; Torres, A.; Molinillo, J.M.G.; Gualtieri, S.C.J.; Macias, F.A. Evaluation of the allelopathic potential of leaf, stem and root extracts of *Ocotea pulchella* Nees et Mart. *Chem. Biodiver.* **2016**, *13*, 1058–1067. [CrossRef] [PubMed]
33. Maraschin-Silva, F.; Aquila, M.E.A. Potencial alelopático de espécies nativas na germinação e crescimento inicial de *Lactuca sativa* L. (Asteraceae). *Acta Bot. Brasil.* **2006**, *20*, 61–69. [CrossRef]
34. El-Gawad, A.A.; Elshamy, A.; El Gendy, A.E.N.; Gaara, A.; Assaeed, A. Volatiles profiling, allelopathic activity, and antioxidant potentiality of *Xanthium strumarium* leaves essential oil from Egypt: Evidence from chemometrics analysis. *Molecules* **2019**, *24*, 584. [CrossRef] [PubMed]

separations

Article

Separation and Disinfection of Contagious Aerosols from the Perspective of SARS-CoV-2

J. J. H. Brouwers

Romico Hold A.V.V., 6226 GV Maastricht, The Netherlands; j.j.h.brouwers@gmail.com

Abstract: An assessment was performed on methods of separating and disinfecting airborne droplet nuclei containing viruses, such as SARS-CoV-2. The droplet nuclei originate from evaporating aerosols emitted by the coughing, singing, sneezing, etc. of infected humans. Based on empirical data and theoretical analysis, we successively determined: (i) the particle number distribution of nuclei versus the nucleus diameter, (ii) the statistical distribution of the viral content in the droplet nuclei starting from a uniform random distribution of viruses in the mucus, (iii) the particle number distribution of droplet nuclei containing at least one virus particle, and (iv) the effectiveness of methods for removing and disinfecting nuclei containing one or more virus particles from indoor air; viz., ventilation with fresh air, filtering with porous media, such as HEPA, and centrifugal separation and simultaneous disinfection, particularly with a rotational particle separator (RPS). Indoor aerosol removal with RPS supplemented with HEPA to arrest tiny volumes of very small particles was found to be the most effective. It is as good as particle removal through ventilation with clean air over long periods of time. An RPS facilitates direct elimination of viruses in the collected nuclei by flushing with a disinfection liquid. The components of an RPS are recyclable. Combining HEPA with an RPS extends the service time of HEPA by almost two orders of magnitude compared to the relatively short service time of stand-alone HEPA filters.

Keywords: SARS-CoV-2; virus particles; air filtering; aerosols; probability distributions; rotational particle separator

Citation: Brouwers, J.J.H. Separation and Disinfection of Contagious Aerosols from the Perspective of SARS-CoV-2. *Separations* **2021**, *8*, 190. https://doi.org/10.3390/separations8100190

Academic Editor: Alena Kubatova

Received: 8 September 2021
Accepted: 11 October 2021
Published: 15 October 2021

1. Introduction

Several epidemic illnesses are caused by viruses [1]. COVID-19, for instance, is caused by infection with the severe acute respiratory syndrome coronavirus 2 (SARS-CoV-2) strain [2]. Transmission of viral diseases, such as COVID-19, may be caused by respiratory droplets in a gas from coughs and/or sneezes. COVID-19 is a new disease, and many of the details of its spread are under investigation at present. This is one of the reasons for why prior literature related to methods and devices configured for decreasing the spread of the SARS-CoV-2 virus is sparse. COVID-19 may spread from person to person via several different modes, particularly when people are in close proximity to one another [3]. Spread may occur very easily and sustainably, with one infected person possibly infecting more than one other person. It may be transmitted when people are symptomatic, but also when people do not show any or only minor symptoms. The World Health Organization (WHO) recommends at least one meter of social distance to hinder transmission, in combination with wearing face masks and washing hands, but people increasingly have problems with this behavior.

Viral diseases in particular spread by means of small droplets in the breath, which are usually referred to as aerosols or bio-aerosols [4–10]. An aerosol is a suspension of fine solid particles or liquid droplets in a gas or in air. The liquid or solid particles can reach very small sizes. As an example, SARS-CoV-2 virus particles are generally about 0.05 μm in size, but they are contained in larger saliva particles. Bacteria are typically 1 μm in size or somewhat larger. The saliva particles and bacteria form an aerosol in the air. When a person emits breath by coughing, sneezing, and the like, the virus-containing aerosol particles are emitted into the air, where they evaporate rather easily, such as in about 1 s, to yield smaller droplet nuclei of about 1 μm that carry the virus particles [8–10]. Animals may also produce such virus-containing droplet nuclei, as well as droplets containing bacteria. The problem with droplet nuclei is that the infectious viruses contained in them may float for quite a long time over quite some distance, thereby increasing the risk for transmission of the disease associated with them.

For instance, it can be shown that, under the influence of gravity, SARS-CoV-2-containing droplet nuclei of about 1 μm will fall in the air for only about 10 cm after 1 h of floating. Larger particles do not have this problem, and particles of, e.g., 10 μm in size will fall in the air at least a hundred times faster. When the virus-containing droplets eventually fall onto floors or surfaces, they may remain infectious if people touch the contaminated surfaces and then their eyes, nose, or mouth with unwashed hands.

Airborne transmission of lower-size aerosols occurs particularly in crowded and less-ventilated indoor spaces [3], which are particularly effective for transmitting viruses, such as in restaurants, nightclubs, public transport, and gatherings such as funerals. Transmission can also occur in a healthcare setting, as certain medical procedures performed on COVID-19 patients generate aerosols. It would be highly desirable to be able to prevent viruses from spreading, particularly in such potentially crowded and less-ventilated indoor spaces.

The aim of this article is to assess methods of separating and disinfecting contagious aerosols containing viruses, particularly the SARS-COVID-2 virus. First, we determine the probability distributions of numbers of droplet nuclei and the distributions of those that can cause infection as a function of their diameter in Sections 2 and 3. The distributions are subsequently used to evaluate the performance of risk mitigation methods in Section 4. Conclusions are presented in Section 5.

2. Probability Density Function of Airborne Droplet Nuclei

The generation of a large number of aerosol droplets through coughing, sneezing, speaking, and breathing has been documented for a long time [4–10]. The droplets are created within the respiratory tract due to disruption of saliva by abrupt changes in the airflow. The front of the mouth is the site of narrowing and the most important site for atomization, with the larynx as second site [7]. The published sizes of these aerosols range from sub-micron sizes to >100 μm, and these can be continuous [8], bi-modal [11], and tri-modal [12]. Care must be taken with these results, however, as the conditions, instruments, and methods differ [13].

When studying bioaerosols generated by human subjects, it is important to distinguish between the initial diameter of particles and the diameter after evaporation of water in ambient air; the resulting desiccated particles are termed "droplet nuclei" in the medical literature. Droplet evaporation is a combination of heat and mass transfer between droplets and their surroundings [14]

$$\frac{\mathrm{d}m_p}{\mathrm{d}t} = -\frac{2\pi p d_p M_v \mathbb{D}\mathrm{Sh}}{RT_\infty} \ln\left(\frac{p - p_{vs}}{p - p_{v\infty}}\right) \tag{1}$$

$$m_p c_p \frac{\mathrm{d}T_p}{\mathrm{d}t} = \pi d_p^2 k_a \frac{T_\infty - T_p}{\frac{1}{2} d_p} \mathrm{Nu} - (h_g - h_l) \frac{\mathrm{d}m_p}{\mathrm{d}t} \tag{2}$$

We estimated the evaporation time by assuming:

- initial droplet temperature $T_p(t = 0) = 37$ °C;
- zero temperature gradient inside the droplet;
- creeping flow around the droplet (Stokes regime).

The physical properties of water and humid air were taken from the Coolprop C++ library [15]. The time constant for evaporation is defined by (see Figure 1):

$$\frac{d_p(t=\tau)}{d_p(t=0)} = \exp\left(-\frac{t=\tau}{\tau}\right) = 0.37 \tag{3}$$

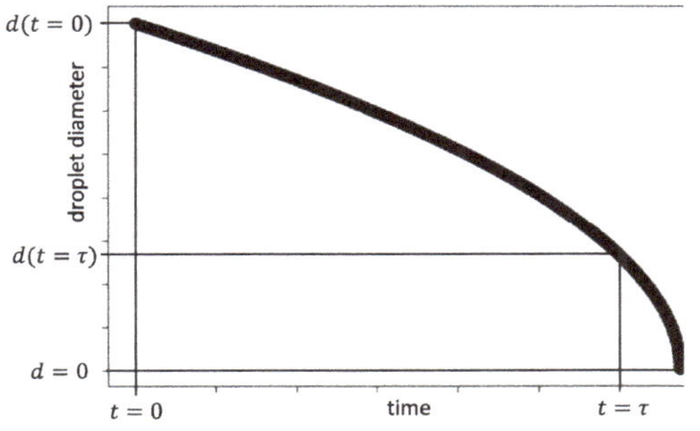

Figure 1. The diameter of an evaporating droplet as a function of time. The time constant is defined at the time at which the droplet is reduced to 0.37 times its original value.

For particles with a diameter <25 μm, the water is evaporated within seconds, even for high values of relative humidity; see Figure 2. These results conform with those of more detailed models [16,17].

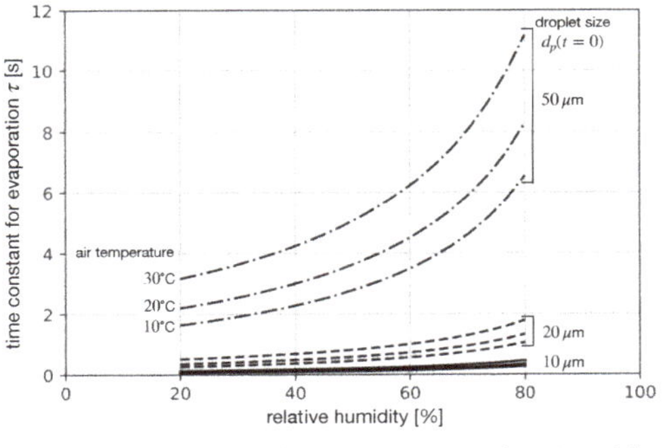

Figure 2. Time constant for evaporation as a function of the relative humidity and temperature of the air.

The amount by which a droplet shrinks upon dehydration depends on the fraction of nonvolatile matter in the oral fluid, which includes electrolytes, sugars, enzymes, DNA, and remnants of dehydrated epithelial and white blood cells [18,19]. As described in [19], whereas pure saliva contains 99.5% water when exiting the salivary glands, the weight fraction of nonvolatile matter in oral fluid falls in the 1 to 5% range. Presumably, this wide range results from differential degrees of dehydration of the oral cavity during normal breathing and speaking and from decreased salivary gland activity with age [19]. Given a nonvolatile weight fraction in the 1 to 5% range and an assumed density of 1.3 g/mL for that fraction, dehydration causes the diameter of an emitted droplet to be reduced in size by a factor of 3–5; see Table 1. As smaller particles are more likely to remain airborne, we use a conservative value of 5 for the shrinkage in the diameter of the emitted droplets.

Table 1. Reduction in the diameter of emitted droplets due to evaporation.

References	Reduction
Duguid (1946) [20]	3–5
Liu et al. (2017) [18]	3
Basu et al. (2020) [17]	3–5
Jarvis (2020) [13]	4

The settling velocity of the resulting droplet nuclei in still air can be calculated by using the Stokes law [21]:

$$v_s = \frac{|\rho_p - \rho_a| g d_d^2}{18\mu_a} \tag{4}$$

wherein ρ denotes the density, g denotes the gravitational acceleration, d denotes the diameter, and μ denotes the dynamic viscosity. The indices p and a denote the droplet nuclei and ambient air, respectively. The settling velocity for water as a function of the droplet diameter in still air is given in Figure 3 . Only droplets with a diameter smaller than 20 μm and with a settling velocity lower than 1 cm/s have a chance of remaining airborne.

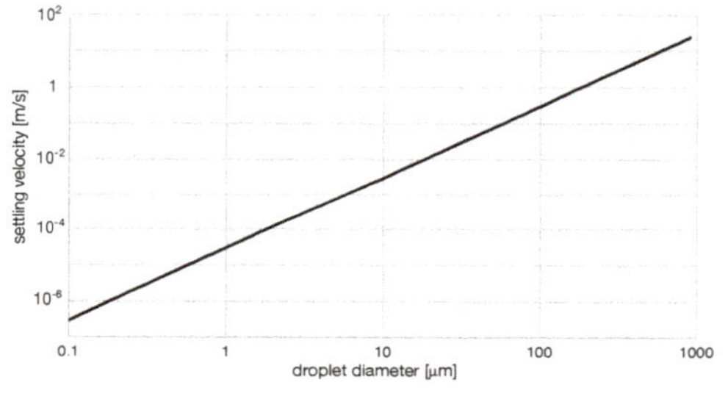

Figure 3. Settling velocity of water droplets in still air as a function of the diameter.

Amongst others, the authors of [9,10,20] investigated the size of droplet nuclei after the evaporation of water; see Figure 4. The emissions from sneezing, coughing, and speaking were investigated. Although the particle number concentration varies for the different modes, the particle number probability density function $f(d_p)$ of the droplet nuclei is similar and follows a log-normal [21] probability density function:

$$f(d_p) = \frac{1}{d_p \ln \sigma_g \sqrt{2\pi}} \exp\left[-\frac{(\ln d_p - \ln \mu)^2}{2(\ln \sigma_g)^2}\right] \tag{5}$$

with a mean of around $\mu = 1$ μm and a geometric standard deviation of approximately $\sigma_g = 2$.

In time, the droplet size may increase due to coagulation. Coagulation is the process whereby aerosol particles collide with one another due to the relative motion between them and adhere to form larger particles [21]. When the relative motion is Brownian, the process is called thermal coagulation. For monodisperse thermal coagulation, we can derive that the change in particle number concentration N is:

$$\frac{\mathrm{d}N}{\mathrm{d}t} = -KN^2 \tag{6}$$

where K is the coagulation constant:

$$K = 4\pi d_p \mathbb{D}_p \tag{7}$$

The Stokes–Einstein equation for the aerosol particle diffusion coefficient is

$$\mathbb{D}_p = \frac{kTC_c}{3\pi\mu d_p} \tag{8}$$

where k denotes the Boltzmann constant, T is the absolute temperature, and μ is the dynamic viscosity. The Cunningham slip correction factor for accounting for deviations from the continuum when calculating the drag of aerosols is:

$$\begin{gathered} C_c = 1 + \frac{2\lambda}{d_d}\left(A_1 + A_2 \exp\frac{-A_3 d_d}{\lambda}\right) \\ A_1 = 1.257\ A_2 = 0.400\ A_3 = 1.10 \end{gathered} \tag{9}$$

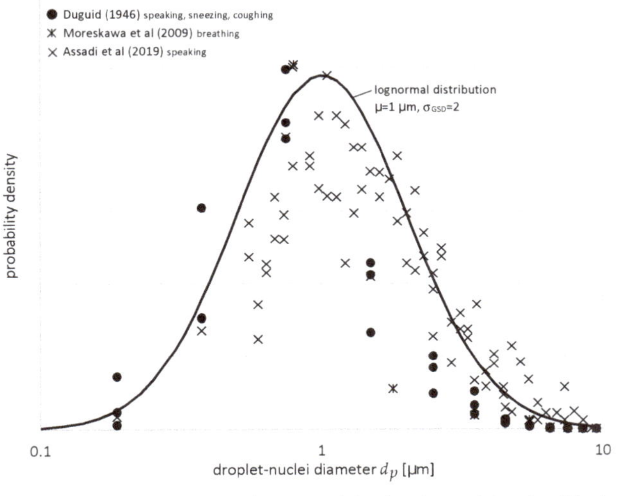

Figure 4. Probability density function of the droplet nuclei emitted by humans.

For air at ambient conditions and a droplet-nucleus diameter of 1 μm, we calculate a coagulation constant of $K = 3.13 \times 10^{-16}$ m^3 s^{-1}. We can determine the particle number concentration as a function of time by integrating Equation (6), assuming K to be constant. The results are shown in Figure 5. It is expected that for droplet nuclei exhaled by humans, coagulation is not significant in normally ventilated rooms.

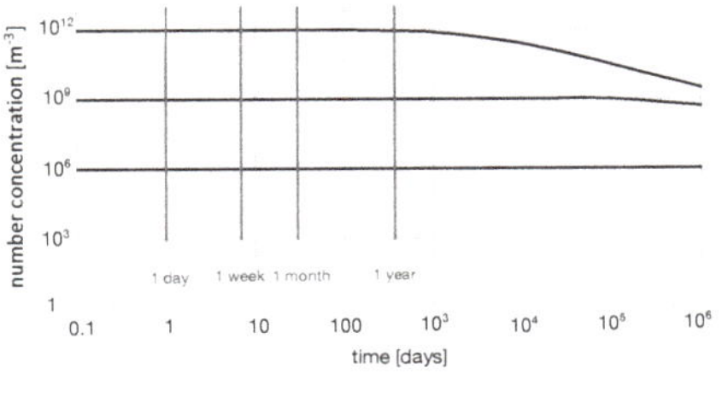

Figure 5. Particle number concentration versus time for simple monodisperse coagulation.

In summary, we conclude that the probability density function of the numbers of airborne droplet nuclei produced by human exhalation is adequately represented by a log-normal distribution with a mean diameter of $\mu = 1$ μm and a geometric standard deviation of $\sigma_g = 2$. Droplet nuclei with a final size > 10 µm, equivalent to an original droplet size of >50 µm, drop out of the air and are deposited on available surfaces.

3. Probability Density Distribution of Numbers of Droplet Nuclei that Can Infect

Having specified the density distribution of droplet nuclei, the question arises of how many of these droplets contain virus particles and can cause a risk of infection. A widely used tool for quantifying the airborne infection risk is the Wells–Riley model [22–24]. It is assumed that there exists a critical dose of bacteria or viruses for which the probability of infection is 63%. The critical dose has to be established through epidemiological research. Subsequently, standard Poisson statistics are used to determine when a person has inhaled this critical dose.

In the vast majority of pathosystems, however, it is unknown how many individual pathogens cause an infection in a host organism. The independent action hypothesis [25] states that each individual pathogen has a non-zero probability of causing host death and that individual pathogens act independently. For viruses, there is no framework for predicting how many virus particles have caused an infection. On the other hand, there is ample experimental evidence that the number of virus particles causing infection can be small for hosts that are highly susceptible [26]. As a conservative approach, we adopt the independent action hypothesis in this work, i.e., even one virus in a droplet nucleus may cause infection of the host.

So far, we have established that pathogens can occur in droplet nuclei with aerodynamic diameters down to their smallest dimensions. What remains is establishing the viral load emitted by a person as a function of the particle probability density function.

The front of the mouth is a site of narrowing and the most important site for atomization, with the larynx as the second site [7]. The concentration of virus particles in the front of the mouth is representative of the concentration in the droplets emitted during the expiratory activities. It is proposed that the droplets emitted by an infected subject have the same viral load c_v, expressed as virus particles per unit volume, as that in the front of the mouth [23]. If we further assume that all of the pathogens that are originally present in the evaporating droplets survive, the viral load is related to the volume of the final droplet nucleus. For our calculations, we use SARS-CoV-2 (COVID-19) as a representative of very small pathogens.

The reported average concentration of SARS-CoV-2 particles in the mouth varies from 10^4 to 10^6 virus particles per milliliter, with extremes between 10^8 and 10^{11} mL^{-1} [27–32]. The Dutch National Institute for Public Health and the Environment (RIVM) [30] fitted the data of Zou et al. [31] and their own data to a normalized log-normal probability density function; see Figure 6. The observed shift towards higher concentrations of the RIVM data was confirmed by another Dutch study (Kampen et al. [32]). To assess the effectiveness of infection risk mitigation measures, we adopt the log-normal probability density function based on the RIVM data in this study. The characteristics of this function are given in Table 2. A cumulative distribution of $F_{c_v} = 0.5$ means that 50% of the patients have a higher concentration and 50% have a lower concentration of virus particles. The mean virus particle concentration is found at this value. To assess the effects of risk mitigation measures, we also use the concentrations at $F_{c_v} = 0.95$ and $F_{c_v} = 0.99$.

Table 2. Characteristics of the log-normal probability density function $f(c_v)$ for the virus particle volume concentration in the mouth based on the RIVM data [30].

Average concentration	μ	7.8×10^{11}	m^{-3}		
Geometric standard deviation	σ_g	20			
Cumulative probability	$F(c_v)$	0.5	0.95	0.99	
Virus particle concentration	c_v	7.8×10^{11}	1.0×10^{14}	7.6×10^{14}	m^{-3}

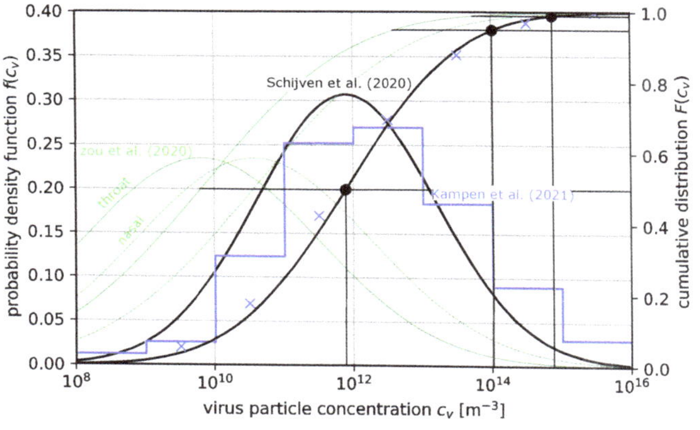

Figure 6. Probability density function and cumulative distribution $F(c_v)$ as a function of the virus particle concentration c_v. Log-normal distributions, adapted from Zou et al. [31] (green) and RIVM data (black), adapted from Schijven et al. [30]. Blue: adapted from Kampen et al. [32].

Now, consider a volume V of fluid in the mouth containing N virus particles. The volume V is expelled into a large number of aerosols, with each aerosol having a volume $v(d_p)$ containing n point-like virus particles; see Figure 7. The probability of finding a virus particle within an aerosol is $p = v/V$, while the probability of finding it in a volume $V - v$ is $q = 1 - p$ [33]. If the particles can be distinguished from each other, the probability of finding any one configuration would be equal to the product of the respective probabilities, $p^n q^{N-n}$. As the virus particles are identical to each other, this number is multiplied by the number of possibilities of choosing n particles within a total of N identical particles, obtaining:

$$p(n) = \left(1 - \frac{v}{V}\right)^{N-n} \left(\frac{v}{V}\right)^n \frac{N!}{(N-n)!n!} \tag{10}$$

This is a binomial distribution that expresses the probability of finding n virus particles in the droplet volume v and the remaining virus particles $N - n$ in the total volume of the other expelled droplets $V - v$.

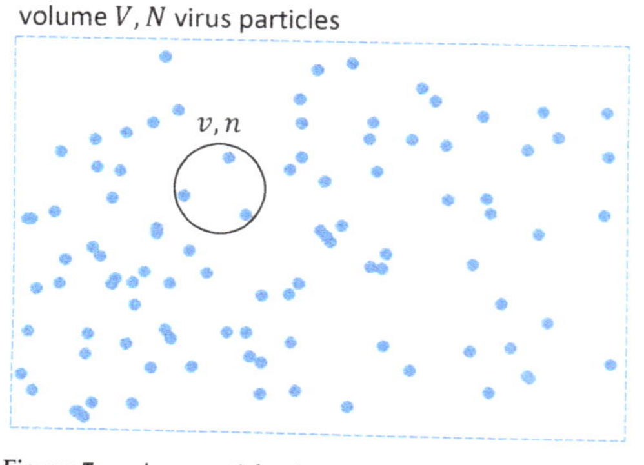

Figure 7. n virus particles in aerosols with volume v inside the total volume of all expelled aerosols V with N randomly dispersed virus particles.

The volume of liquid V in the mouth from which aerosols are generated can be assumed to be large in comparison with the volume v of the individual aerosols. Under these circumstances, the binomial distribution can be approximated with the Poisson distribution (Larsen [34]). With a relative error of v/V, one can show:

$$p(n) = \frac{(c_v v)^n \exp(-c_v v)}{n!} \tag{11}$$

In accordance with the independent action hypothesis, we conservatively assume that just one virus in a nucleus is already sufficient to cause the infection of a host. That is, all cases with $n > 0$ lead to infection. The probability of this happening is given by

$$p_{n>0} = 1 - p_{n=0} = 1 - \exp(-c_v v) \tag{12}$$

Noting that in accordance with the analysis of the previous section, the diameter of the original aerosol emitted from the mouth is five times the diameter of the resulting droplet nucleus, we have

$$v = \frac{\pi}{6}(5d_p)^3 \tag{13}$$

where, as before, d_p is the diameter of the droplet nucleus. The distribution now becomes

$$p_{n>0}(\overline{d_p}) = 1 - \exp\left(-N_d \overline{d_p}^3\right) \tag{14}$$

where $\overline{d_p}$ is the droplet diameter divided by its mean diameter μ:

$$\overline{d_p} = d_p/\mu \tag{15}$$

and

$$N_d = \frac{\pi}{6}(5\mu)^3 c_v \tag{16}$$

is the number of virus particles in a droplet nucleus of mean diameter μ. From solution (14), it is seen that $p_{n>0}(\overline{d_p})$ grows from 0 as $N_d\overline{d_p}^3 << 1$ goes to 1 as $N_d\overline{d_p}^3 >> 1$. As we found that $\mu = 1$ µm, the value of N_d is equal to 5.1×10^{-5}, 6.5×10^{-3}, or 5.0×10^{-2} for the values of c_v from Table 2 of 7.8×10^{11}, 1.0×10^{14}, and 7.6×10^{14} m^{-3}, respectively. A small value of N_d implies that the value of $p_{n>0}(\overline{d_p})$ gets a value that is appreciably larger than zero for only relatively large values of $\overline{d_p}$; more precisely, for the values of $N_d\overline{d_p}^3 \approx 1$ or $\overline{d_p} \approx N_d^{-1/3}$. For the quoted values of c_v and N_d, this implies values of d_p of about 27, 5.4, and 2.7 µm, respectively. In summary, the nuclei corresponding to the left-hand side of the probability density distribution of the nuclei have a lower chance of infecting. This feature is illustrated in Figure 8, where the probability density distribution of numbers of nuclei that can infect versus their diameter is shown for the quoted values of virus particles in the mouth liquid. This probability density is defined by the original density of droplet nuclei—cf. Equation (5)—multiplied by the probability of having at least one virus in the droplet nucleus, according to Equation (14). This is given by

$$f'(\overline{d_p}) = a_0^{-1}\overline{d_p}^{-1}\left[\exp\left(-\frac{1}{2}\left(\frac{\ln \overline{d_p}}{\ln \sigma_g}\right)^2\right)\right]\left[1 - \exp\left(-N_d\overline{d_p}^3\right)\right] \tag{17}$$

where

$$a_0 = \int_0^\infty \overline{d_p}^{-1}\left[\exp\left(-\frac{1}{2}\left(\frac{\ln \overline{d_p}}{\ln \sigma_g}\right)^2\right)\right]\left[1 - \exp\left(-N_d\overline{d_p}^3\right)\right] d\overline{d_p} \tag{18}$$

is a normalization constant that ensures that the surface underneath the distribution is unity. With increasing viral concentration, the small droplet nuclei pick up more virus particles so that the most probable value of the density distribution shifts to lower values of the nucleus diameters; see Figure 8. This feature was already noticed in 1946 by Duguid [20].

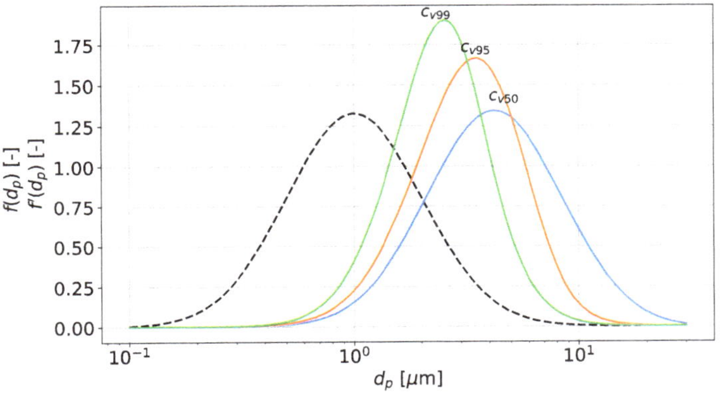

Figure 8. Probability density $f(d_p)$ of droplet nuclei (broken line) and probability density $f'(d_p)$ of droplet nuclei containing at least one virus particle for $\sigma_g = 2$ and $\mu = 1$ μm. Viral concentrations in the mouth liquid: $c_{v50} = 7.8 \times 10^{11}$, $c_{v95} = 1.0 \times 10^{14}$, and $c_{v99} = 7.6 \times 10^{14}$ m^{-3}.

4. Methods of Risk Reduction

An effective way to remove infectious aerosols from indoor air is to ventilate with clean air from outside. Denoting ventilation flow by ϕ and the volume of the room or space that is ventilated by V_o, the decrease in the number of particles inside the space or room as a function of time t can be described by the ordinary differential equation

$$V_o \frac{dN_{n>0}}{dt} = -\phi N_{n>0} \tag{19}$$

where $N_{n>0} = N_{n>0}(t, d_p)$ is the number of droplet nuclei of diameter d_p containing at least one virus particle per unit volume of air. The initial condition imposed on the differential equation is:

$$N_{n>0}(t, d_p) = N_{n>0}(0) f'(d_p) \qquad at \qquad t = 0 \tag{20}$$

where $N_{n>0}(0)$ is the initial value of the total number of droplet nuclei with at least one virus particle per unit volume of air and $f'(d_p)$ is the particle number distribution of these nuclei as a function of diameter. The initial particle number distribution is assumed to correspond to that generated by an infected person or persons and is given by Equation (17). Spatial variations in the numbers of droplet nuclei inside the room that is ventilated have been disregarded. Such variations will depend on, among other things, the positions of particle sources and of inlets and outlets of vent air [35]. Studying their effects is beyond the scope of the present analysis. The aim is to define a case based on clean air that can be used as a reference to determine the effectiveness of removing infectious particles with filtering methods.

Noting that the integral of $f'(d_p)$ taken over all diameters of droplet nuclei with at least one virus particle is unity, the solution of Equation (19) for the total number of droplet nuclei with at least one virus particle $N_{n>0}(t)$ divided by the initial total number is

$$\frac{N_{n>0}(t)}{N_{n>0}(0)} = e^{-t/\tau_c} \tag{21}$$

where

$$\tau_c = V_o/\phi \tag{22}$$

is the time needed to replace the air in the room with clean outside air once. The inverse of τ_c is known as air changes per hour (ACH) [36]. A typical value for τ_c in residential buildings is 15 to 30 min. The concentration will then have been reduced by exp(-2) to exp(-4), that is, by a factor of 1/7 to 1/55 in one hour.

4.1. Filtering with Porous Media

For cases where ventilation with clean outside air is not feasible or insufficient, filtering of inside air is an option. A common configuration is a fan that generates flow through a filter comprising a porous medium within which particles are entrapped. The previous formulation for ventilation with outside air can be directly extended to one where the air is filtered by replacing ϕ with $E_f(d_p)\phi$:

$$\phi \longrightarrow E_f(d_p)\phi \tag{23}$$

In this case, ϕ is the flow through the filter device and $E_f(d_p)$ is the collection efficiency of the filter, i.e., the number of particles at an inlet of a given diameter minus the number at an outlet of the same diameter divided by the number at the inlet of the same diameter. In general, the collection efficiency will vary with the particle diameter. Overall efficiency follows from the relation

$$E = \frac{\int_0^\infty N_{n>0}(t,d_p)E_f(d_p)d(d_p)}{\int_0^\infty N_{n>0}(t,d_p)d(d_p)} \tag{24}$$

where E is the collection efficiency of all particles distributed according to $N_{n>0}(t,d_p)$. Filter media that are used in practice, such as high-efficiency particle filters (HEPA), will collect almost all particles with diameters as small as 0.1 µm and even smaller [37]. This implies that nuclei that can infect will be collected with almost 100% efficiency; $E_f = E = 1$. A filter device with flow ϕ will be as good as ventilation with the flow of clean outside air of the same size. Solution (21) also holds for the filter with flow ϕ. If a porous filter medium with a lower efficiency is applied, e.g., with about 95% efficiency, this can be compensated by only 5% extra flow through the device to achieve equivalence in particle number reduction performance, assuming that the collection efficiency is the same for all diameters of nuclei.

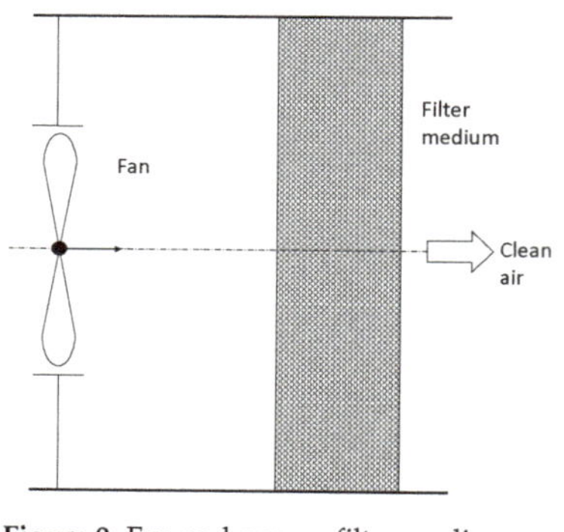

Figure 9. Fan and porous filter medium.

Filter media are widely used in indoor air filtering, from centralized air treatment in multi-story buildings to stand-alone in-house air cleaners; Figure 9. What undoubtedly contributes to their popularity is their simplicity in design. What is not always recognized is their drawback: With time, they become clogged, and the pressure drop over the filter medium increases. For a given ventilator characteristic, the filter flow decreases and, as directly follows from the previous formulae, the performance in concentration reduction drops. In practice, this feature often remains unnoticed until the point where the flow has dropped to almost nothing. To keep performance deterioration to a minimum, filter media must be replaced at regular intervals. Replacement is not risk-free because potentially large numbers of infectious particles are assembled inside the filter media. Furthermore, filter media contain non-recyclable components and can suffer from re-emission and degassing of the deposited particles. These disadvantages can be minimized by combining filter media with a rotational particle separator; see Section 4.3.

4.2. Centrifugal Separation

Clogging and increased reduction of flow with time are absent in centrifugal separation. The main types of centrifugal separators are the vane separator, the cyclone, and the rotational particle separator (RPS); Figure 10 [38]. These three methods are the most well known from applications in industry. The cyclone separator was introduced in the domestic market as a vacuum cleaner by Dyson [39]. A compelling argument is that the consumer no longer needs filter bags. They reduce the suction power over time and have to be regularly replaced. The RPS was invented in 1987 [40] and was followed by a number of spin-offs, the last of which was patented in May 2021 [41].

The basis of centrifugal separation is that, in the case of flow of a gas that makes a curve with heavier particles inside it, particles will be directed away from the gas stream lines because of the action of the centrifugal force. The separation is completed through the arrival of the particles at a collecting wall. The formulae that enable the calculation of the separation performance of the three methods are presented in Figure 10.

The separation performance of a centrifugal separator can, in general, be summarized with a single function of the separation efficiency versus particle diameter made non-dimensionless by $d_{p50\%}$; see Figure 11. The parameter $d_{p50\%}$ corresponds to the particle diameter for which all particles with a larger diameter have a collection efficiency larger than 50%. The value of $d_{p50\%}$ is determined by the design parameters of the centrifugal separator (Figure 10).

The overall collection efficiency of droplet nuclei with at least one virus particle follows from relation (24). Here, the value of the integral will depend on the value of $d_{p50\%}$. The result is shown in Figure 12 for the case of the separation of droplet nuclei that can infect according to Equation (17). The calculations are based on the theoretical separation curve in Figure 11.

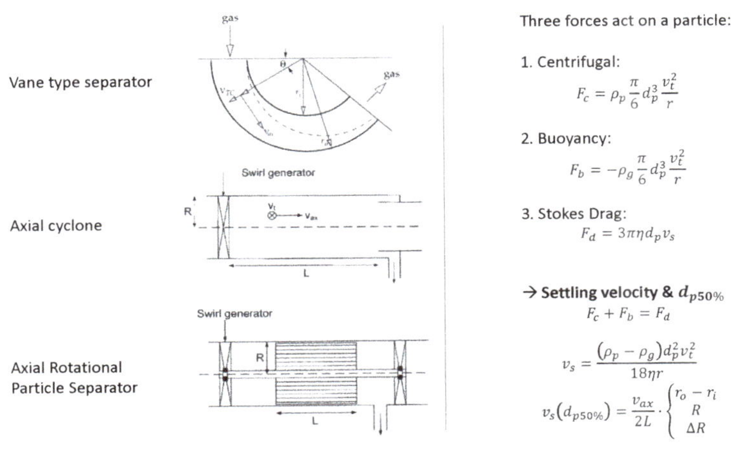

Figure 10. Centrifugal separation methods.

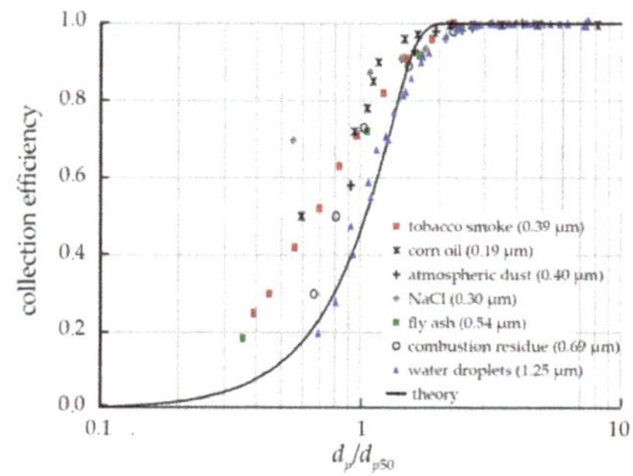

Figure 11. Efficiency with which particles are separated versus particle diameter divided by d_{p50} measured for a range of RPS designs that are subjected to different particle loads. The values of $d_{p50\%}$ of the various designs are indicated between brackets. Next to the measured results is the theoretically derived full line for the RPS. This line represents an approximate description of the efficiency of cyclone- and vane-type separators, reprinted with permission from ref. [38].

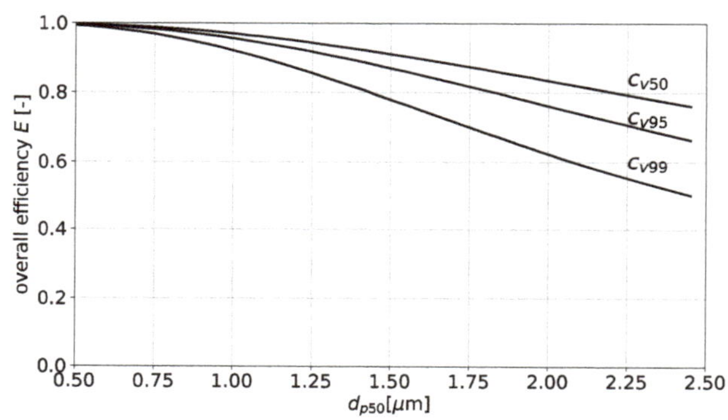

Figure 12. Degree to which particles distributed according to Equation (17) are separated by the centrifugal method as a function of the design parameter d_{p50} for various viral concentrations in the mucus.

Apart from separation performance, practical designs are dictated by requirements on size and energy consumption. For the vane separator and cyclone, the values of $d_{p50\%}$ are typically 10 µm and larger and 4 µm and larger, respectively. As seen from Figure 12, this makes them unsuited for nucleus separation. Practical versions of the RPS, on the other hand, have values of $d_{p50\%}$ of around 0.3 µm and larger. They enable the separation of particles down into the sub-micron range. The separation performance has been tested for a variety of RPS designs that were subjected to different kinds of particle loads; Figure 11. The results were published in a number of papers that were reviewed in [38]. The measured separations were consistently found to correspond to theoretical predictions based on solutions of the classical laws of mechanics. They can be represented by the curve in Figure 11. The RPS technology is a candidate for effective separation of droplet nuclei.

4.3. Rotational Particle Separator (RPS)

As follows from Figure 12, the efficiency of the removal of almost 100% of viruses was obtained for the design values of $d_{p50\%}$ of 1 µm and less. Such design values of $d_{p50\%}$ are common for practical RPS designs; Table 3.

It should be noted, however, that the separation degrees shown in Figure 12 apply to the initial situation, where the distribution of droplet nuclei versus the diameter of those that can infect corresponds to the values given by the initial condition (20). With time, the larger nuclei will first be removed. The smaller ones, however, are subjected to lower filter efficiency and take more time to be filtered. To deal with this situation, we need to solve Equation (19) with ϕ replaced by $E_f(d_p)\phi$ and noting that $E_f(d_p)$ varies with the nucleus diameter. One obtains

$$\frac{N_{n>0}(t)}{N_{n>0}(0)} = \int_0^\infty f'(d_p)e^{-E_f(d_p)t/\tau}d(d_p) \tag{25}$$

In Figure 13, the rate of decrease in the number of nuclei that can infect versus time has been shown for various cases. The reference is the decrease in the case of ventilation with clean outside air. It is seen that with time, the rate of decrease in the case of RPS is reduced. With time, the larger nuclei that are collected with 100% efficiency are directly taken out of the air after passing through the RPS. The very small ones, whose particle number concentration is small, need to pass through the RPS several times in order to be removed completely. In the case of an RPS with $d_{p50\%} = 1$ µm, the reduction rate starts to deteriorate when the numbers of nuclei have reached a value of about 10^{-2} times the concentration at the start. The deterioration of the removal rate can be delayed by applying a lower value of $d_{p50\%}$. For example, in the case of an RPS with $d_{p50\%} = 0.5$ µm, the rate of reduction starts to deteriorate when the numbers of nuclei have reached a value of about 10^{-4} times the concentration at the start. The deterioration can be eliminated altogether by introducing a HEPA filter at the outlet of the RPS. Consider, for example, an RPS with $d_{p50\%}$ =1 µm. It will remove about 95% of all particles that can infect right from the start, as well as all other airborne particles in the room. In this case, the HEPA filter is subjected to a minimal load of about 2% of the volume of all particles that are not collected by the RPS. The service life of the HEPA filter will be practically two orders of magnitude longer compared to that without the RPS.

Table 3. Example of the design of a rotational particle separator.

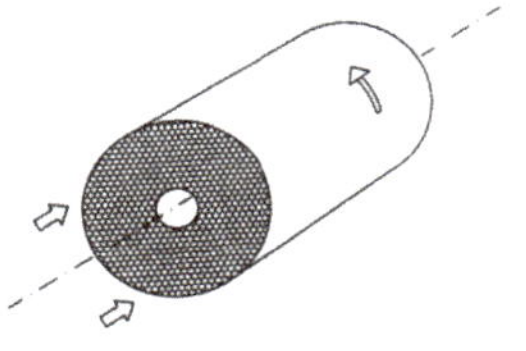

Parameter	Symbol	Example Value
Filter flow	ϕ_t	0.1 m^3/s
Channel diameter	$d_c(\Delta R)$	3×10^{-3} m
Angular velocity	Ω	200 rad/s
Length of cylinder	L	0.2 m
Radius of cylinder	R	0.15 m
Viscosity of air	η	1.8×10^{-5} kg/m×s
Density of particles	ρ_p	10^3 kg/m^3
Density of air	ρ_g	1.3 kg/m^3
-	$d_{p50\%} = \left(\frac{27\eta\phi_t d_c}{2(\rho_p-\rho_g)\Omega^2 L\pi R^3}\right)^{1/2}$	0.92 µm
Pressure drop over filter	$\Delta p_t = \frac{64\eta\phi_t}{\pi d_c^2 R_o^2}$	180 Pa

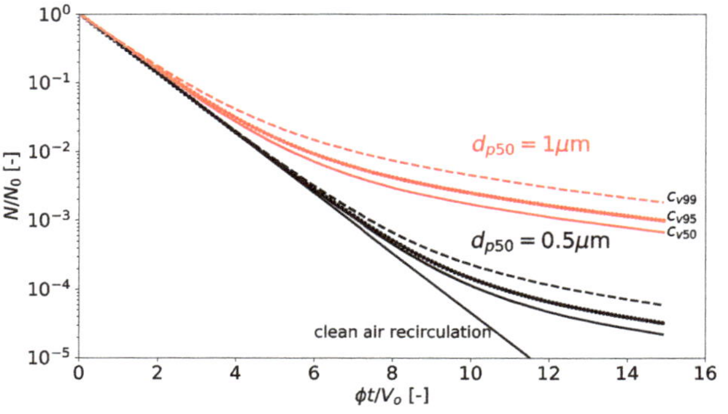

Figure 13. Rate of decrease in the number of nuclei that can infect versus dimensionless time. Comparison of the performance of the RPS with $d_{p50\%} = 1$ µm and $d_{p50\%} = 0.5$ µm and with clean air ventilation for various concentrations of virus particles in the mucus. The RPS supplemented with HEPA follows the curve of clean air recirculation.

4.4. RPS with Liquid Flushing

The separation by the RPS described in the previous section bears the risk of re-entrainment of particles collected at the radially outward sides of the separation channels. This risk is particularly present in periods in which the rotation rate is reduced, whereby the centrifugal force that presses the particles to the walls fades. Moreover, the de-gassing of the particles can cause the emission of unwanted components into the filtered air. The way out is to apply liquid flushing, a technique that has been tested and applied in de-misting versions of the RPS [38] and the more recently conceived rotating absorber device [41].

Liquid is sprayed at the top of the vertically positioned rotating channel structure. The liquid enters the channels and, due to the centrifugal force, quickly forms a liquid film at the outer walls. It flows downwards as a thin film. Particles entrained in the gas are centrifugated to the outer sides of the channel walls, as described in a quantitative manner in the previous section. Here, they enter the liquid film and flow with the liquid to the bottom of the rotating channel structure; Figure 14. Liquid leaves the channels at the bottom as large droplets that contain the particles are propelled to a radially positioned collection chamber. The liquid collection chamber is of a limited volume—a few liters in total—and serves as a reservoir whose content is re-injected at the top.

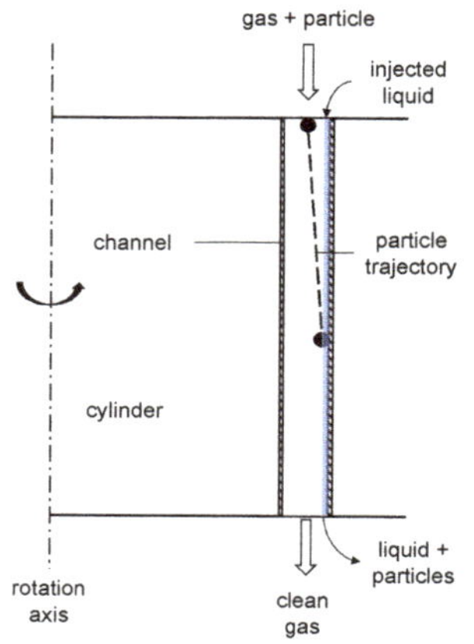

Figure 14. Liquid flushing in a vertical RPS.

The pressure drop and separation performance of the RPS supplemented with liquid flushing remain constant over time. The liquid needs to be replaced only after relatively long periods of time, which can be up to a year. Moreover, the liquid preferably consists of disinfecting components that immediately kill all viruses and bacteria. The RPS technology consists of recyclable components and fits into modern circular economics. The combination of the RPS with liquid flushing and the possible supplementation with a HEPA filter positioned at the gas outlet of the device offers particle separation performance that is equal to that of clean air ventilation.

5. Conclusions

The size of aerosols emitted from the mucus of humans by coughing, sneezing, singing, speaking, etc. has been measured and documented extensively. Due to evaporation shortly after emission, the aerosols are reduced in diameter, on average, by a factor of 3 to 5. The resulting diameter distributions of the droplet nuclei can be captured with a log-normal distribution with a mean diameter of 1 μm and a geometric standard deviation of 2.

Given a uniform random distribution of virus particles in the mucus, the probability of finding one or more virus particles in an aerosol or nucleus can be described by a Poisson distribution. For common viral concentrations in the mucus, only larger-sized nuclei contain at least one virus particle. The diameter above which one or more virus particles are present decreases with the viral concentration. Combining the probability of having one or more virus particles with the probability density of the droplet nuclei yields the distribution of numbers of droplet nuclei that contain at least one virus particle as a function of the nucleus diameter. The parameter in these distributions is the viral concentration in the mucus, for which the latest data of RIVM were used.

To assess the effectiveness of indoor aerosol removal methods, ventilation with clean outside air was used as a reference. Well-designed filters based on porous media, such as HEPA, remove almost all potentially infectious particles. The drawback is that, with time, the filter media become saturated and clogged. The pressure drop increases, and the flow through the filter device becomes less and less. The rate by which the number of particles in the room that can infect are reduced deteriorates. The filter media have to be regularly replaced and disposed of as hazardous chemical waste.

Filter bags are absent in the case of separation devices based on centrifugal methods. The flow through the devices remains constant, as there is no increase in the pressure drop with time. From the options of centrifugal methods based on vane-type separators, cyclones, and rotational particle separators (RPSs), only an RPS is capable of removing particles with diameters going into the sub-micron range typical of droplet nuclei that contain virus particles with sufficient efficiency. In an RPS, the flow through the device remains undisturbed. All particles that are collected are continuously removed by flushing with a liquid, which is then recirculated. By applying a disinfecting liquid, all virus particles are killed directly after being collected inside the channels of the device. The device remains virus-free, and its components are based on sustainable technology.

The collection efficiency of the RPS drops below 100% for nuclei with diameters that are substantially less than 1 μm. The result is that when performing filtering in a room, the nuclei with very small sub-micron diameters need to be recirculated several times before being removed from the contaminated air in the room to a sufficient degree. The particle number concentration of these very small sub-micron-size nuclei is low, but their presence leads to a slower rate of nucleus reduction after some time of filtering, i.e., when the concentration has reached a level of 10^{-2} to 10^{-4} times the concentration at the start of filtering. To prevent this from happening, an HEPA filter can be installed at the gas outlet of the RPS. This ensures a constant rate of reduction of all potentially infectious nuclei, which is equal to that of ventilation with clean outside air at an equal air flow. The time until the replacement of the HEPA filter can be almost two orders of magnitude longer than that of HEPA as a stand-alone filter.

Particle separation with an RPS and HEPA, liquid flushing in the RPS and disinfection by liquids are established methods of practice. The effective elimination of contagious aerosols in rooms will be demonstrated in the future.

Funding: This research received no external funding.

Data Availability Statement: Not applicable.

Acknowledgments: H.P. van Kemenade is acknowledged for his contribution to the contents of Sections 2 and 3 in particular. B.G.J. Ruis is acknowledged for the preparation of this article.

Conflicts of Interest: The authors declare no conflict of interest.

References

1. Wikipedia. Coronavirus Diseases. Available online: https://en.wikipedia.org/wiki/Coronavirus_diseases (accessed on 1 August 2021).
2. Wikipedia. COVID-19. Available online: https://en.wikipedia.org/wiki/COVID-19 (accessed on 1 August 2021).
3. Qian, H.; Miao, T.; Liu, L.; Zheng, X.; Luo, D.; Li, Y. Indoor transmission of SARS-CoV-2. *Indoor Air* **2021**, *31*, 639–645. [CrossRef] [PubMed]
4. Hutchison, R.F. Die verbreitung von keimen durch gewöhnliche luftströme. *Z. FÜR Hyg. Und Infekt.* **1901**, *36*, 233. [CrossRef]
5. Wells, W.F. On air-borne infection. Study II. Droplets and droplet nuclei. *Am. J. Hyg.* **1934**, *20*, 611–18.
6. Tellier, R. Aerosol transmission of influenza A virus: A review of new studies. *J. R. Soc. Interface* **2009**, *6*, 783–790. [CrossRef] [PubMed]
7. Johnson, G.R. Modality of human expired aerosol size distributions. *J. Aerosol Sci.* **2011**, *42*, 839–851. [CrossRef]
8. Han, Z.; Weng, W.; Huang, Q. Characterizations of particle size distribution of the droplets exhaled by sneeze. *J. R. Soc. Interface* **2013**, *9*, 2013056. [CrossRef] [PubMed]
9. Morawska, L. Size distribution and sites of origin of droplets expelled from the human respiratory tract during expiratory activities. *J. Aerosol Sci.* **2009**, *40*, 256–269. [CrossRef]
10. Assadi, S. Aerosol emission and superemission during human speech increase with voice loudness. *Sci. Rep.* **2019**, *9*, 2348. [CrossRef]
11. Bao, L.; Gao, H.; Deng, W.; Lv, Q.; Yu, H.; Liu, M.; Yu, P.; Liu, J.; Qu, Y.; Gong, S.; et al. Transmission of severe acute respiratory syndrome coronavirus 2 via close contact and respiratory droplets among human angiotensin-converting enzyme 2 mice. *J. Infect. Dis.* **2020**, *222*, 551–555. [CrossRef]
12. Gralton, J.; Tovey, E.; McLaws, M.L.; Rawlinson, W.D. The role of particle size in aerosolised pathogen transmission: A review. *J. Infect.* **2011**, *62*, 1–13. [CrossRef]
13. Jarvis, M.C. Aerosol transmission of SARS-CoV-2: Physical principles and implications. *Front. Public Health* **2020**, *8*, 813. [CrossRef]
14. Bird, R.B.; Stewart, W.E.; Lightfoot, E.N. *Transport Phenomena*; John Wiley & Sons: New York, NY, USA, 1960.
15. Bell, I.; Wronski, J.; Quolin, S. Coolprop C++ Library. Available online: http://coolprop.org/ (accessed on 7 February 2021).
16. Liu, F.; Qian, H.; Zheng, X.; Song, J.; Cao, G.; Liu, Z. Evaporation and dispersion of exhaled droplets in stratified environment. In *IOP Conference Series: Materials Science and Engineering*; IOP Publishing: 2019; Volume 609-4, p. 042059.
17. Basu, S.; Kabi, P.; Chaudhuri, S.; Saha, A. Insights on drying and precipitation dynamics of respiratory droplets from the perspective of COVID-19. *Phys. Fluids* **2020**, *32*, 123317. [CrossRef]
18. Liu, L.; Wei, J.; Li, Y.; Ooi, A. Evaporation and dispersion of respiratory droplets from coughing. *Indoor Air* **2017**, *27*, 179–190. [CrossRef]
19. Stadnytskyi, V.; Bax, C.E.; Bax, A.; Anfinrud, P. The airborne lifetime of small speech droplets and their potential importance in SARS-CoV-2 transmission. *Proc. Natl. Acad. Sci. USA* **2020**, *117*, 11875–11877. [CrossRef] [PubMed]
20. Duguid, J. The size and the duration of air-carriage of respiratory droplets and droplet-nuclei. *Epidemiol. Infect.* **1946**, *44*, 471–479. [CrossRef]
21. Hinds, W.C. *Aerosol Technology: Properties, Behavior, and Measurement of Airborne Particles*; John Wiley & Sons: Hoboken, NJ, USA, 1999.
22. Riley, E.; Murphy, G.; Riley, R. Airborne spread of measles in a suburban elementary school. *Am. J. Epidemiol.* **1978**, *107*, 421–432. [CrossRef]
23. Buonanno, G.; Stabile, L.; Morawska, L. Estimation of airborne viral emission: Quanta emission rate of SARS-CoV-2 for infection risk assessment. *Environ. Int.* **2020**, *141*, 105794. [CrossRef] [PubMed]
24. Guo, Y.; Qian, H.; Sun, Z.; Cao, J.; Liu, F.; Luo, X.; Ling, R.; Weschler, L.B.; Mo, J.; Zhang, Y. Assessing and controlling infection risk with Wells-Riley model and spatial flow impact factor (SFIF). *Sustain. Cities Soc.* **2021**, *67*, 102719. [CrossRef]
25. Druett, H. Bacterial invasion. *Nature* **1952**, *170*, 288–288. [CrossRef] [PubMed]
26. Zwart, M.P. An experimental test of the independent action hypothesis in virus–insect pathosystems. *Proc. R. Soc. B Biol. Sci.* **2009**, *276*, 2233–2242. [CrossRef]

27. Wölfel, R. Virological assessment of hospitalized patients with COVID-2019. *Nature* **2020**, *581*, 465–469. [CrossRef]
28. Fajnzylber, J.; Regan, J.; Coxen, K.; Corry, H.; Wong, C.; Rosenthal, A.; Worrall, D.; Giguel, F.; Piechocka-Trocha, A.; Atyeo, C.; et al. SARS-CoV-2 viral load is associated with increased disease severity and mortality. *Nat. Commun.* **2020**, *11*, 1–9. [CrossRef] [PubMed]
29. Kleiboeker, S.; Cowden, S.; Grantham, J.; Nutt, J.; Tyler, A.; Berg, A.; Altrich, M. SARS-CoV-2 viral load assessment in respiratory samples. *J. Clin. Virol.* **2020**, *129*, 104439. [CrossRef]
30. Schijven, J.F.; Vermeulen, L.C.; Swart, A.; Meijer, A.; Duizer, E.; de Roda Husman, A.M. Exposure assessment for airborne transmission of SARS-CoV-2 via breathing, speaking, coughing and sneezing. *medRxiv* **2020**.
31. Zou, L.; Ruan, F.; Huang, M.; Liang, L.; Huang, H.; Hong, Z.; Yu, J.; Kang, M.; Song, Y.; Xia, J.; et al. SARS-CoV-2 viral load in upper respiratory specimens of infected patients. *N. Engl. J. Med.* **2020**, *382*, 1177–1179. [CrossRef]
32. van Kampen, J.J.; van de Vijver, D.A.; Fraaij, P.L.; Haagmans, B.L.; Lamers, M.M.; Okba, N.; van den Akker, J.P.; Endeman, H.; Gommers, D.A.; Cornelissen, J.J.; et al. Duration and key determinants of infectious virus shedding in hospitalized patients with coronavirus disease-2019 (COVID-19). *Nat. Commun.* **2021**, *12*, 1–6. [CrossRef] [PubMed]
33. Mauri, R. *Non-Equilibrium Thermodynamics in Multiphase Flows*; Springer: Berlin/Heidelberg, Germany, 2012.
34. Larsen, M.L. Spatial distributions of aerosol particles: Investigation of the Poisson assumption. *J. Aerosol Sci.* **2007**, *38*, 807–822. [CrossRef]
35. Blocken, B.; van Druenen, T.; Ricci, A.; Kang, L.; van Hooff, T.; Qin, P.; Xia, L.; Ruiz, C.A.; Arts, J.; Diepens, J.; Maas, G.; Gillmeier, S.; Vos, S.; Brombacher, A. Ventilation and air cleaning to limit aerosol particle concentrations in a gym during the COVID-19 pandemic. *Build. Environ.* **2021**, *193*, 107659. [CrossRef] [PubMed]
36. Wikipedia. Ventilation (Architecture). Available online: https://en.wikipedia.org/wiki/Ventilation_(architecture) (accessed on 1 August 2021).
37. Wikipedia. HEPA. Available online: https://en.wikipedia.org/wiki/HEPA (accessed on 1 August 2021).
38. van Kemenade, E.; Brouwers, B.; van Benthum, R. Centrifugal separation with emphasis on the rotational particle separator. *ChemBioEng Rev.* **2014**, *1*, 262–272. [CrossRef]
39. Wikipedia. James Dyson. Available online: https://en.wikipedia.org/wiki/James_Dyson (accessed on 27 August 2021).
40. Brouwers, J.J.H. Rotational Particle Separator. European Patent EP0286160B1, 12 October 1988.
41. Brouwers, J.J.H. Rotational Absorber Device and Method for Scrubbing an Absorbate from a Gas. European Patent EP3624923B1, 14 April 2021.

separations

MDPI

Article

Simultaneous Determination of Seven Bioactive Constituents from *Salvia miltiorrhiza* in Rat Plasma by HPLC-MS/MS: Application to a Comparative Pharmacokinetic Study

Yanli Zhang [1,†], Weiliang Cui [2,†], Xianghong Liu [1], Ning Wang [3], Wenru Kong [3], Junyu Sui [4], Huifen Li [5,*] and Shuqi Wang [3,*]

1 Department of Pharmacy, Qilu Hospital of Shandong University, Cheeloo College of Medicine, Jinan 250012, China; yanlizhangqlyy@163.com (Y.Z.); liuxianghong666@126.com (X.L.)
2 Shandong Institute for Food and Drug Control, Jinan 250101, China; xinyao2002@126.com
3 School of Pharmaceutical Sciences, Shandong University, Jinan 250012, China; wangningsdyxy@163.com (N.W.); qfkongwenru@163.com (W.K.)
4 College of Agriculture & Life Science, University of Wisconsin, Madison, WI 53706, USA; jsui3@wisc.edu
5 College of Pharmacy, Shandong University of Traditional Chinese Medicine, Jinan 250355, China
* Correspondence: lhfen007@126.com (H.L.); wangsq@sdu.edu.cn (S.W.); Tel.: +86-0531-8962-8200 (H.L.); +86-0531-8838-2014 (S.W.)
† Both these authors contribute equally to this work.

Citation: Zhang, Y.; Cui, W.; Liu, X.; Wang, N.; Kong, W.; Sui, J.; Li, H.; Wang, S. Simultaneous Determination of Seven Bioactive Constituents from *Salvia miltiorrhiza* in Rat Plasma by HPLC-MS/MS: Application to a Comparative Pharmacokinetic Study. *Separations* **2021**, *8*, 93. https://doi.org/10.3390/separations8070093

Academic Editor: Victoria Samanidou

Received: 30 May 2021
Accepted: 25 June 2021
Published: 29 June 2021

Abstract: The roots of *Salvia miltiorrhiza* (Danshen) is a precious herbal medicine used to treat cardiovascular diseases. This study establishes a high-performance liquid chromatography-tandem mass spectrometric (HPLC-MS/MS) method to quantify seven bioactive constituents from Danshen in rat plasma simultaneously. Chromatographic separation is performed on an Agilent Eclipse Plus C18 column (150 mm × 2.1 mm, 5 μm), utilizing a gradient of acetonitrile and 0.2% formic acid aqueous solution as the mobile phase, at a flow rate of 0.6 mL/min. We conduct a tandem mass spectrometric detection with electrospray ionization (ESI) interface via multiple reaction monitoring (MRM) in both positive and negative ionization mode. Our results show that a linear relationship is established for each analyte of interest over the concentration range of 0.5–300 ng/mL with $r \geq 0.9976$. The validated method is successfully used to compare the pharmacokinetic properties of crude and wine-processed Danshen extract orally administered to rats. C_{max} of tanshinone IIA, C_{max}, and $AUC_{0\text{-}t}$ of dihydrotanshinone I decrease significantly ($p < 0.05$) in the wine-processed group. No significant changes for other compounds are observed. These results might provide meaningful information for the further application of wine-processed Danshen and understanding of wine-processing mechanisms.

Keywords: *Salvia miltiorrhiza*; HPLC-MS/MS; pharmacokinetics; wine-processed

1. Introduction

Bioactive components of the roots of *Salvia miltiorrhiza* (Danshan) have been developed into various formulations clinically to treat microcirculatory disturbance-related diseases, such as heart disease, chronic hepatitis, diabetes, early cirrhosis, cerebral ischemia, and cancer [1–3]. Chemical investigation of Danshen in the past few years has revealed that two major types of secondary metabolites were responsible for its therapeutic effects: lipophilic tanshinones and hydrophilic phenolic acids [4,5]. These components exhibited multiple biological activities via different mechanisms. For example, tanshinone IIA and dihydrotanshinone I could enhance autophagy and induce proteasomal degradation of the Tau protein, resulting in increased Amyloid-β clearance and decreased Tau phosphorylation, making them potential candidates for AD treatment in the future [6,7]. Cryptotanshinone and tanshinone I exhibit vigorous antiviral activities against SARS-CoV-2 with IC_{50} values of 5.63 and 2.21 μmol/L, respectively [8]. Salvianolic acids, rosmarinic acid, caffeic acid, ferulic acid, and lithospermic acid are four representative hydrophilic phenolic acids isolated from Danshen [9,10]. Among them, salvianolic acid A can reduce cardiotoxicity induced by arsenic trioxide through decreasing cardiac mitochondrial injury and shows great potential against cancer cells via targeting various signaling pathways [11,12]. Rosmarinic acid, ferulic acid, and lithospermic acid play several biological roles, including free radical scavenger, inhibitor of prooxidant enzymes that catalyze free radical production, and enhancer of scavenger enzyme [13–15].

Pharmacokinetic description of the bioactive components in Danshen will provide scientific evidence regarding their properties of adsorption, distribution, metabolism, and excretion in vivo. Understanding the pharmacokinetic properties of each component enables prescribers to choose appropriate dose and dose intervals to ensure the safety and efficiency of drug application. HPLC–MS/MS method has been extensively applied to determine the pharmacokinetic profiles of different bioactive components from Danshen because of its high sensitivity and specificity [16–18]. An efficient UPLC/MS/MS method that employs isocratic elution and positive/negative ionization switching analysis, has been established for the determination of four salvianolic acids and four tanshinones in Danshen simultaneously in 2 min [19]. However, several aspects still need to be improved for their application in biological samples. Higher sensitivity and efficiency, as well as a larger concentration range for the quantification, remain to be achieved by optimizing the HPLC-MS/MS and sample pretreatment conditions.

Wine-processing is a traditional method for treating herbal medicine before clinical use to achieve multiple purposes: modify the taste, reduce the toxicity, and enhance the biological activity [20,21]. Wine-processed Danshen has been reported to possess enhanced blood—quickening and stasis—transforming, as well as antimicrobial activities. Although the crude and wine-processed Danshen exhibited different chemical profiles, detailed exploration of the chemical basis behind medicinal property changes after the processing has not been performed yet [22–24]. Moreover, a comparison between pharmacokinetic properties of crude and wine-processed Danshen in vivo, which is meaningful for their clinical reasonable application and understanding the wine-processing mechanism, has not been characterized yet.

In the current work, a sensitive and efficient method for simultaneous quantification of seven bioactive constituents from Danshen in plasma samples was developed and applied to compare the pharmacokinetic properties of crude and wine-processed Danshen extract orally administered to rats. Hydrophobic tanshinones, including tanshinone IIA, cryptotanshinine, and dihydrotanshinone I, as well as hydrophilic phenolic acids, including salvianolic acid A, rosmarinic acid, ferulic acid, and lithospermic acid (Figure 1), were simultaneously determined for the first time at a concentration range of 0.5–300 ng/mL with $r \geq 0.9976$ in 8 min. Pharmacokinetic parameters, such as AUC, C_{max}, and $t_{1/2}$, were determined and compared. This will provide information about the influence of wine-processing on the pharmacokinetics, as well as the pharmacological activity of Danshen bioactive components.

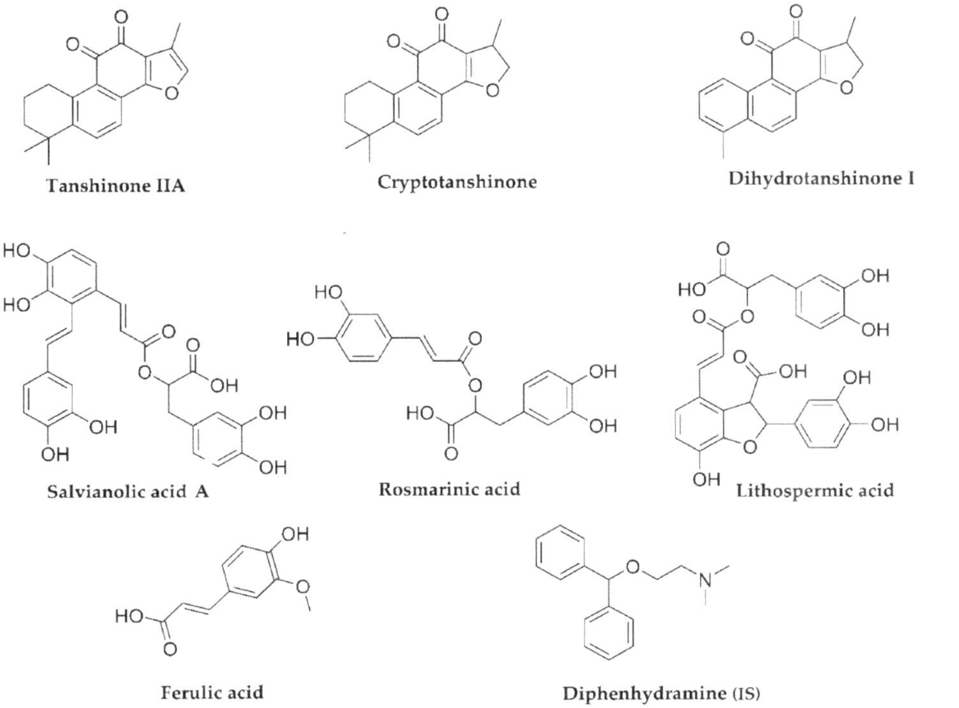

Figure 1. Chemical structure of seven components and internal standards (IS): Tanshinone IIA; cryptotanshinone; dihydrotanshinone I; salvianolic acid A; rosmarinic acid; lithospermic acid; ferulic acid and diphenhydramine (IS).

2. Materials and Methods

2.1. Chemical Reagents

Danshen were collected from Jinan Green Chinese Herbal Pieces Co., (Jinan, Shandong, China) and authenticated by Professor Xu Lingchuan in the field. The wine-processed Danshen were collected from Shandong Jianlian Shengjia Traditional Chinese Medicine Co. (Jinan, Shandong, China) and authenticated by Professor Xu Lingchuan. Tanshinone IIA (Lot: 110766-200416), cryptotanshinone (Lot: 110852-201307), rosmarinic acid (Lot: 11871-201303), ferulic acid (Lot: 110773-200608), and diphenhydramine (Lot: 100066-200807) were purchased from China National Institutes for Food and Drug Control. Salvianolic acid A (Lot: B20260), lithospermic acid (Lot: B21683), and dihydrotanshinone I (Lot: B20357) were purchased from Shanghai Yuanye Biotechnology Co. Ultra-pure water and chemicals, such as acetonitrile and formic acid of analytical grade purity, were used in the experiment.

2.2. Instruments, Liquid Chromatography, and Mass Spectrometry Conditions

A Nanospace SI-2 HPLC system (Shiseido, Japan) equipped with a NASCA 5100 autosampler, a vacuum degasser unit, and a binary pump, and an Agilent C18 column (150 mm × 2.1 mm, 5 μm) were selected for chromatographic analysis. An API 5500 Q-Trap triple quadrupole mass spectrometer (AB SCIEX, Concord, ON, Canada) equipped with a TurboIonSpray source was used for mass detection. Data were processed on Analyst 1.5.2 software package. The solvent flow was diverted from the MS after the first minute of the gradient. Samples were ionized using an electrospray ion (ESI) source in both positive and negative mode. The ionization voltage for positive and negative mode was +4.5 kV and −4.0 kV, respectively. The source temperature was set at 55 °C. Nitrogen was used as the curtain gas (35 psi), nebulizer gas (GS1, 55 psi), and turbo gas (GS2, 55 psi).

2.3. Preparation of Calibration Standards, Internal Standard (IS), and Quality Contro (QC) Samples

To prepare the stock solution of calibration standards and QC samples, all compounds were dissolved in acetonitrile to a final concentration of 1 mg/mL. The stock solutions were serially diluted with acetonitrile to prepare the working solutions of calibration standards and QC samples. The calibration standards were prepared by spiking a specific volume of the working solutions into the corresponding biological samples to yield a final concentration of 0.5, 1, 3, 5, 10, 30, 50, 100, and 300 ng/mL, respectively. Diphenhydramine at 400 ng/mL was used as the IS working solution. The low, medium, and high levels of QC samples containing analytes at the concentration of 0.5, 5, and 240 ng/mL were prepared in the same manner. All the solutions were stored at 4 °C for further use.

2.4. Preparation of Crude and Wine-Processed Danshen Extract

Extracts were prepared according to the following procedures. Powder of crude and wine-processed Danshen were firstly extracted twice by heating reflux at 80 °C for 2 h in 80% ethanol, and the extract was concentrated with rotary evaporation under vacuum at 60 °C. The residual was then extracted twice with water by heating reflux at 80 °C for 2 h and concentrated in the same manner. Extract obtained after the four extractions were combined and dried, and then crushed into powder for further experiments. The contents of tanshinone IIA, cryptotanshinone, dihydrotanshinone I, salvianolic acid A, rosmarinic acid, ferulic acid, and lithospermic acid were 28.3 ± 0.08, 19.6 ± 0.22, 64.0 ± 1.36, 217.0 ± 2.32, 505.0 ± 4.91, 14.9 ± 0.22, 1240.0 ± 13.5 ng/mg in crude Danshen extract, and 33.5 ± 1.13, 23.1 ± 1.07, 72.3 ± 1.39, 166.0 ± 3.56, 564.0 ± 4.97, 16.5 ± 0.43, 1010.0 ± 4.32 ng/mg in wine-processed Danshen extract, respectively.

2.5. Preparation and Handling of Biological Samples

An aliquot of 10 μL IS working solution (400 ng/mL) and 200 μL acetonitrile were added into a 20 μL plasma sample. The mixture was vortex mixed for 5 min to precipitate the proteins and then centrifuged at 14,000 rpm for 10 min at 4 °C. The supernatant was transferred to a 1.5 mL Eppendorf tube, and 5 μL of the supernatant was injected into the HPLC-MS/MS system for analysis.

2.6. Method Validation

The HPLC-MS/MS method was developed and validated according to the Guidance for Industry Bioanalytical Method Validation [25]. The full validation, including selectivity, linearity, accuracy, precision, matrix effect, extraction recovery, and stability, was carried out in the plasma matrix.

2.6.1. Selectivity

To evaluate the selectivity of the method, the chromatogram of blank plasma samples from six different lots of rats was compared with those of corresponding plasma samples spiked with a standard solution of the seven analytes, and plasma samples after oral administration of Danshen extract.

2.6.2. Linearity and Lower Limit of Quantification (LLOQ)

The final concentration of calibration standards for plotting the calibration curve was 0.5, 1, 3, 5, 10, 30, 50, 100, and 300 ng/mL. The curve was plotted with the mass concentration of each drug in the plasma on the abscissa (X), and the peak area ratio of the drug to the IS on the ordinate (Y). The weighting factor $1/x^2$ was used for the best fit line of $y = kx + c$ using linear regression analysis. The correlation coefficient (r) of 0.995 or better was considered as the best response for quantification analysis. The concentration of analytes in the QCs or test samples was calculated based on the regression parameters obtained from the calibration curves. The analyte response at the LLOQ should be at least 10 times of blank response.

2.6.3. Accuracy, Precision, and Recovery

Intra-day precision and accuracy were estimated at three different QC levels, i.e., 0.5, 5, and 240 ng/mL, by analyzing six replicates in a single day. The inter-day precision was determined by analyzing the three different QC samples on nine different runs in three consecutive days. Relative error (RE) and related standard deviation (RSD) were used to evaluate the accuracy and the precision, respectively. The mean values for RE and RSD should be within 15% of the actual value except at LLOQ, where it should not deviate from the mean value by more than 20%.

The recovery of each analyte at three different concentrations was evaluated by comparing the relative peak area of the analyte spiked in the rat plasma to that of the standard directly dissolved in solvent at the same concentration. The ratio gives the recovery.

2.6.4. Stability and Matrix Effect

The stability of the analytes was evaluated in triplicates at three different QC levels: 0.5 ng/mL, 5 ng/mL, and 240 ng/mL. A freeze-thaw stability experiment was performed by subjecting the QC samples to three freeze and thaw cycles from −80 °C to room temperature. Samples were left overnight in an autosampler setting at 4 °C or at room temperature for 4 h to evaluate their autosampler stability and benchtop stability. Bias was calculated against the freshly prepared QC samples. Samples with a difference within ±15% were considered stable. The matrix effect on the quantification of analytes was evaluated by comparing the peak area ratio of the analyte spiked in plasma samples to that of the standard directly dissolved in solvent at the same concentration.

2.7. Application to a Pharmacokinetic Study

Male Wistar rats (220–230 g, n = 12) used in the experiments were supplied by the Lab Animal Center of Shandong University (Grade II, Certificate No. SYXK 2013-0001). Rats fasted for 12 h before drug administration and for a further 2 h after dosing, and have free access to water during experiments. The experimental protocol was approved by the University Ethics Committee and conformed to the "Principles of Laboratory Animal Care" (NIH publication no. 85-23, revised 1985). The rats were randomly divided into two groups (six in each group) and conducted with a single oral dose of crude and wine-processed Danshen extract suspended in water (10 g/kg), respectively.

After oral administration, 150 μL of blood samples were collected from the jugular vein at different time points (before dosing and at 0.25, 0.5, 1.0, 1.5, 2.0, 4.0, 4.5, 6.0, 12.0, 24.0, and 48.0 h post-dosing) and put into heparinized tubes, which were centrifuged at 3000× *g* for 10 min at 4 °C. The supernatant was stored at −80 °C and analyzed using the method described above within one month. The plasma concentration of different drugs at different time points was determined based on the standard curve. The plasma drug concentration was plotted on the ordinate, and time was on the abscissa, yielding the concentration–time curves.

Parameters, including the peak plasma concentration (C_{max}) and time to peak concentration (T_{max}), were obtained from experimental observations. The other pharmacokinetic parameters were analyzed using the program TOPFIT (version 2.0; Thomae GmbH, Germany) according to a non-compartmental model. The linear trapezoidal rule to approximately the last point was used to calculate the area under the plasma concentration–time curve ($AUC_{0\text{-}t}$). Dividing the area under the first moment–time curve ($AUMC_{0\text{-}t}$) by the area under the curve ($AUC_{0\text{-}t}$) yields the mean residence time (*MRT*). Total oral body clearance (*CL/F*) was calculated using the following equation: $CL/F = \text{dose}/AUC_{0\text{-}t}$. All results were expressed as mean ± standard deviation (SD). Statistical comparisons between different groups were analyzed using SPSS version 20.0 (SPSS Inc., Chicago, IL, USA) by an analysis of variance (ANOVA).

3. Results and Discussion

3.1. *Method Development*

3.1.1. Optimization of the HPLC-MS/MS Conditions

Methanol-water or acetonitrile-water as the mobile phase in the gradient elution system was firstly compared in the aspects of retention time, signal intensity, and resolution. We found that the acetonitrile-water mobile phase system yielded chromatographic peaks with better resolution and intensity than methanol-water did. The addition of 0.2% formic acid into the mobile phase significantly increased the signal intensity of the eight compounds and improved their peak shapes. Thus, the acetonitrile (A)/water (B) with 0.2% formic acid was selected as the elution solvent system. The HPLC gradient system was set up as follows—13% of A for 0.5 min, then linearly increased to 90% of A in 2.5 min, and finally a decrease to 13% of A in 0.5 min prior to column re-equilibration. The total running lasted 8 min.

The standard solution of the seven analytes and the IS was separately introduced into the ESI source in either positive or negative ionization mode to determine the pattern of the most abundant ions in these compounds. The four phenolic acids showed an intense signal response and less noise in negative mode, while the three tanshinones and IS provided higher signal intensity in positive mode. Based on these results, we designed a segment-program to detect the eight compounds with the most vital signals in both positive and negative modes. Optimal MS parameters for the MRM scan mode were listed in Table 1.

Table 1. MRM transitions and parameters for detecting tanshinone IIA, cryptotanshinone, dihydrotanshinone I, salvianolic acid A, rosmarinic acid, ferulic acid, lithospermic acid, and diphenhydramine.

Analytes	ESI Mode	Precursor Ion (*m*/*z*)	Product Ion (*m*/*z*)	DP	CE
Tanshinone IIA	positive	295.1	277.2	130	26
Cryptotanshinone	positive	297.1	279.1	100	29
Dihydrotanshinone I	positive	279.1	261.2	200	22
Salvianolic acid A	negative	493.3	295.1	−100	−24
Rosmarinic acid	negative	359.1	293.1	−150	−33
Ferulic acid	negative	193.1	134.1	−100	−26
Lithospermic acid	negative	537.1	493.1	−68	−11
IS	positive	256.2	167.1	200	16

Abbreviations: ESI, electrospray ion; CE, collision energy; DP, declusting potential.

3.1.2. Optimization of Extraction Procedure

Protein precipitation, solid-phase extraction, and liquid phase extraction are three commonly used methods for sample extraction. Protein precipitation using organic solvents has been widely accepted because of its high efficiency, convenience, and low cost [26,27]. In the current study, different organic solvents for protein precipitation were attempted during sample preparation. Finally, acetonitrile was selected because of its excellent efficiency, preferable recovery (87.3~105.6%), and little matrix effect. In the case of methanol, the recovery of rosmarinic acid, and lithospermic acid is relatively poor.

3.2. *Method Validation*

3.2.1. Specificity

Figure 2 shows the typical MRM chromatograms of blank plasma, blank plasma spiked with the seven analytes and IS, and the sample collected from rats at 1h after oral administration of crude Danshen extract. Peaks of all the seven components and IS, at the retention times of 4.89 min (tanshinone IIA), 4.45 min (cryptotanshinone), 4.00 min (dihydrotanshinone I), 2.10 min (salvianolic acid A), 2.09 min (rosmarinic acid), 1.97 min (ferulic acid), 2.09 min (lithospermic acid) and 2.23 min (IS), did not interfere with those from endogenous substances.

3.2.2. Linearity and LLOQ

The linear ranges, regression equations, and correlation coefficients of the seven analytes are shown in Table 2. All calibration curves exhibited good linearity over the concentration range from 0.5 ng/mL to 300 ng/mL with correlation coefficient (r) ranging from 0.9956 to 0.9976. Based on a signal peak-to-noise ratio = 10, the LLOQs for all the eight compounds were 0.5 ng/mL.

Table 2. Linear ranges, regression equations, and correlation coefficients of tanshinone IIA, cryptotanshinone, dihydrotanshinone I, salvianolic acid A, rosmarinic acid, ferulic acid, and lithospermic acid in rat plasma.

Analytes	Concentration Range (ng/mL)	Regression Equation	Correlation Coefficient (r)
Tanshinone IIA	0.5–300	Y = 0.828X + 0.0137	0.997
Cryptotanshinone	0.5–300	Y = 0.125X + 0.0744	0.996
Dihydrotanshinone I	0.5–300	Y = 0.0695X − 0.0205	0.997
Salvianolic acid A	0.5–300	Y = 0.00478X + 0.0137	0.996
Rosmarinic acid	0.5–300	Y = 0.00434X + 0.00434	0.998
Ferulic acid	0.5–300	Y = 0.00332X + 0.00174	0.997
Lithospermic acid	0.5–300	Y = 0.000721X + 0.00376	0.996

3.2.3. Precision, Accuracy

The intra-day and inter-day precisions (RSD) of the seven analytes at three different concentration levels ranged from 2.2% to 9.3%, while the accuracy (RE) of the samples ranged from −2.1% to 6.4% (Table 3). All these values were within the acceptable range, implying that the method was reproducible and reliable.

Table 3. Summary of precision, accuracy, recovery, and matrix effect of tanshinone IIA, cryptotanshinone, dihydrotanshinone I, salvianolic acid A, rosmarinic acid, ferulic acid, and lithospermic acid in rat plasma (n = 9, three consecutive days).

Analytes	Spiked (ng/mL)	Intra-Day Precision RSD (%)	Inter-Day Precision RSD (%)	Intra-Day Accuracy RE (%)	Recovery Mean ± SD (%)	Matrix Effect Mean ± SD (%)
Tanshinone IIA	0.5	6.2	7.4	6.4	87.4 ± 5.0	109.1 ± 3.4
	5.0	5.2	5.8	4.3	89.4 ± 6.5	105.4 ± 4.2
	240	3.4	4.2	4.7	91.7 ± 4.2	110.3 ± 2.5
Cryptotanshinone	0.5	7.8	8.9	2.7	85.3 ± 7.6	102.5 ± 1.7
	5.0	5.9	4.2	−1.5	89.2 ± 3.4	92.3 ± 2.5
	240	4.7	5.2	4.6	91.3 ± 6.1	97.4 ± 6.4
Dihydrotanshinone I	0.5	8.1	6.7	−2.1	89.4 ± 6.9	91.4 ± 7.1
	5.0	3.5	5.5	6.2	91.2 ± 7.5	93.9 ± 2.0
	240	4.3	5.7	5.5	92.6 ± 8.4	98.1 ± 5.5
Salvianolic acid A	0.5	7.2	9.3	3.2	106.7 ± 8.7	96.4 ± 4.1
	5.0	6.8	5.7	0.8	87.9 ± 2.4	106.5 ± 3.4
	240	4.6	3.6	2.9	105.4 ± 4.7	103.7 ± 3.6
Rosmarinic acid	0.5	4.1	5.2	4.5	94.6 ± 4.2	89.4 ± 5.7
	5.0	2.4	2.2	1.4	92.7 ± 5.1	91.6 ± 6.8
	240	3.3	5.8	2.6	102.5 ± 7.8	95.5 ± 1.5
Ferulic acid	0.5	9.2	8.7	5.5	87.3 ± 5.4	112.4 ± 2.4
	5.0	6.5	6.7	6.2	91.6 ± 4.7	104.8 ± 5.1
	240	4.7	5.5	5.4	96.5 ± 6.4	105.4 ± 2.8
Lithospermic acid	0.5	7.1	6.5	2.7	105.6 ± 8.7	92.4 ± 3.4
	5.0	5.6	7.6	5.5	96.8 ± 7.9	89.7 ± 4.8
	240	2.8	4.0	6.2	90.9 ± 6.2	86.7 ± 6.5
IS	400	N.D.	N.D.	N.D.	94.7 ± 5.4	N.D.

3.2.4. Extraction Recoveries and Matrix Effects

The extraction recoveries and matrix effects of the investigated analytes in rat plasma are shown in Table 3. The recovery rates of the investigated analytes ranged from 87.3% to 105.6%, with SD values lower than 8.7%, 7.9%, and 8.4% at three concentration levels, respectively. The matrix effects of the eight analytes of interest varied from 86.7% to 112.4%, with SD values lower than 7.1%, 6.8%, and 6.5% at low, medium, and high concentrations, respectively. These results suggested that the sample processing method for extraction of analytes from biological samples exhibited high efficiency, and the recoveries for all the eight analytes were in acceptable ranges. No significant matrix effect existed.

3.2.5. Stability

All the seven analytes were stable in rat plasma under different experimental conditions: 4 °C for 24 h (RE: −9.6~9.2%, RSD ≤ 9.8%), −80 °C for 30 days (RE: −8.8~9.2%, RSD ≤ 8.9%), and after three freeze thaw cycles at −20 °C (RE: −3.3~7.2%, RSD ≤ 8.6%). All the results were within the acceptance criteria of ±15% deviation from the nominal concentration (Table 4).

Table 4. Stability of tanshinone IIA, cryptotanshinone, dihydrotanshinone I, salvianolic acid A, rosmarinic acid, ferulic acid, and lithospermic acid in rat plasma samples (n = 5).

Analytes	Spiked (ng/mL)	Stability at 4 °C for 24 h (RSD, %)	(RE, %)	Stability at −80 °C for 30 Days (RSD, %)	(RE, %)	Freeze-Thaw Stability (RSD, %)	(RE, %)
Tanshinone IIA	0.5	6.8	4.0	4.1	3.6	6.7	3.1
	5.0	5.9	−2.4	4.8	6.8	5.9	−2.8
	240	6.5	−2.6	1.6	2.0	3.0	4.8
Cryptotanshinone	0.5	8.0	−7.2	8.9	5.6	8.3	2.4
	5.0	3.6	9.2	6.5	6.4	7.0	6.4
	240	3.2	2.4	5.9	−3.7	2.8	1.9
Dihydrotanshinone I	0.5	9.8	−2.4	7.4	−5.2	6.7	−1.2
	5.0	4.5	5.6	4.3	0.8	3.7	4.8
	240	1.9	1.9	4.0	3.2	3.0	4.3
Salvianolic acid A	0.5	5.2	6.0	6.0	1.2	6.4	2.8
	5.0	6.1	2.4	3.6	7.2	6.8	−2.8
	240	1.7	−0.9	4.8	−3.6	4.6	−3.3
Rosmarinic acid	0.5	8.2	−1.2	6.7	−8.8	4.9	−1.2
	5.0	2.9	3.2	5.1	4.8	2.8	7.6
	240	1.3	2.5	3.6	4.1	3.0	4.3
Ferulic acid	0.5	3.7	−9.6	7.6	6.0	8.0	7.2
	5.0	5.3	5.6	5.2	5.6	7.3	4.0
	240	3.2	3.6	6.2	1.5	4.4	3.0
Lithospermic acid	0.5	7.2	4.8	3.5	1.6	4.0	8.4
	5.0	3.5	8.4	4.2	9.2	8.6	3.6
	240	3.0	0.9	5.4	2.3	2.9	3.9

3.3. Pharmacokinetic Study

The above-validated method was successfully applied to a comparative pharmacokinetic study in rats after oral administration of crude and wine-processed Danshen extract at a dose of 10 g/kg. The mean plasma concentration-time profiles of seven analytes are plotted and presented in Figure 3. The main non-compartmental pharmacokinetic parameters are listed in Table 5.

Based on the concentration–time curves and the pharmacokinetic parameters, the effect of wine-processing on the pharmacokinetics of the seven ingredients of Danshen in the rats were examined. As shown in Figure 3, the plasma concentration of all the components reached maximum quickly (T_{max} < 1 h), and wine-processing did not change the T_{max} values significantly. Compared to the crude Danshen group, the C_{max} of all the hydrophobic tanshinones decreased to some extent, and a significant decrease was observed for tanshinone IIA and dihydrotanshinone I ($p < 0.05$) in the wine-processed Danshen group. As to the hydrophilic acids, we did not see any obvious differences in C_{max} between the two groups. In addition, the $AUC_{0\text{-}t}$ and $AUC_{0\text{-}\infty}$ of dihydrotanshinone I decreased significantly; CL/F of salvianolic acid A and dihydrotanshinone I increased significantly in the wine-processed Danshen group.

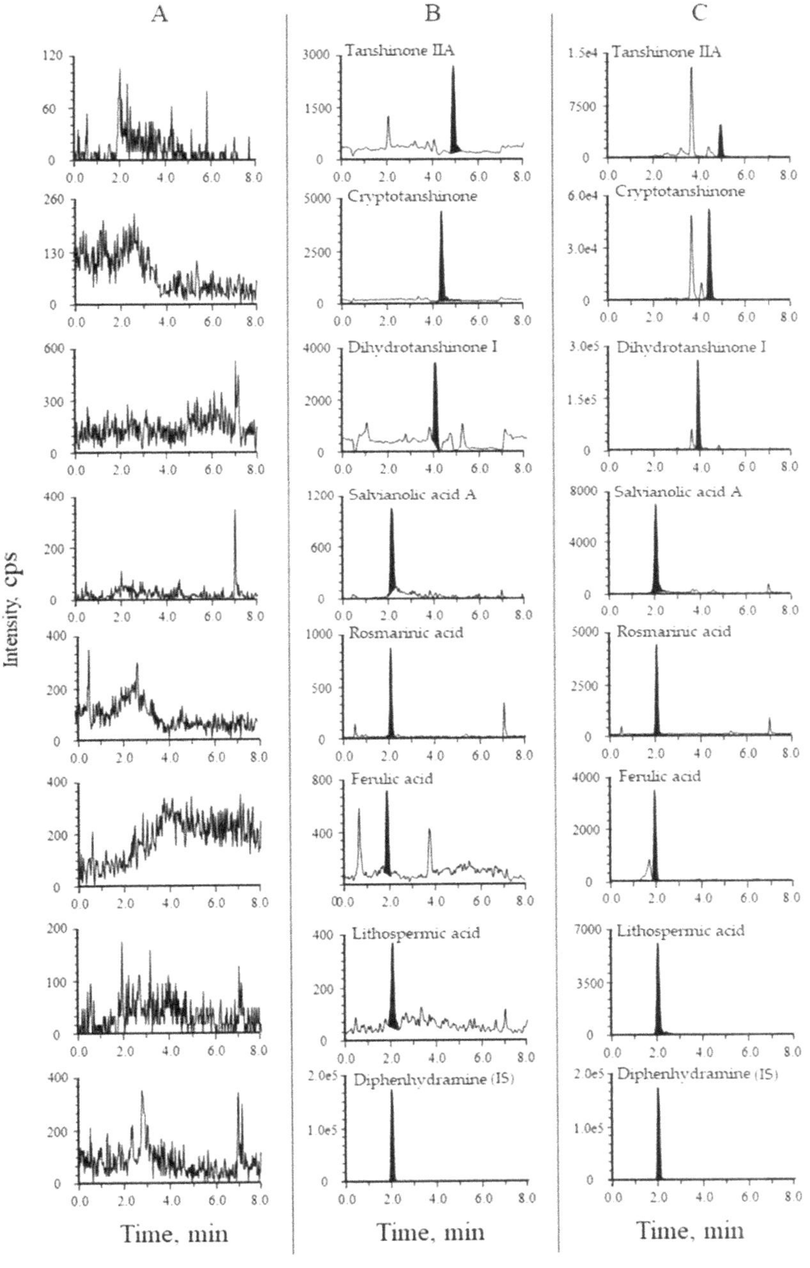

Figure 2. Representative MRM chromatograms of blank rat plasma (**A**); blank plasma spiked with seven analytes and IS (**B**); rat plasma sample collected at 1 h after oral administration of Danshen extract (**C**). Tanshinone IIA, cryptotanshinone, dihydrotanshinone I and IS were detected in positive mode. Salvianolic acid A, rosmarinic acid, ferulic acid, and lithospermic acid were detected in negative mode.

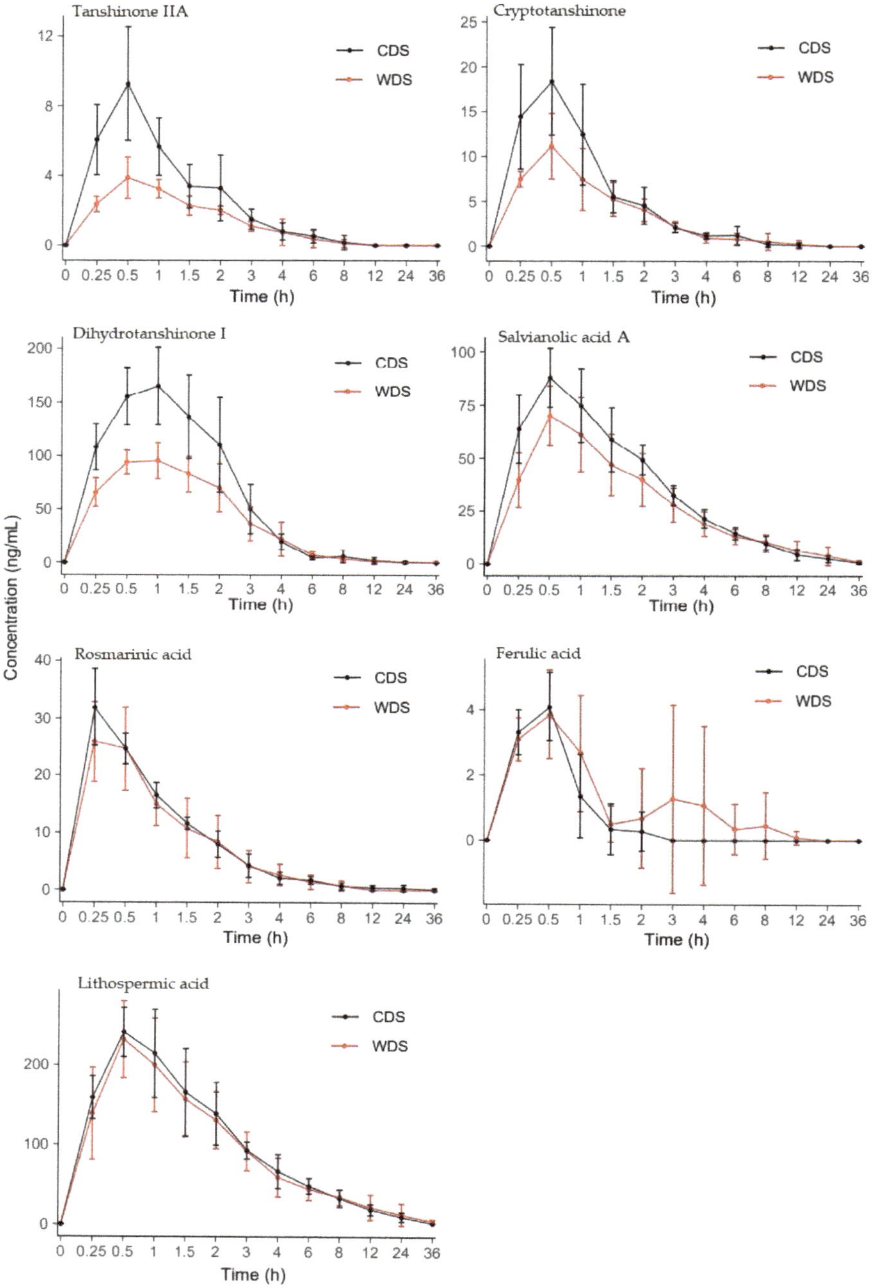

Figure 3. Mean (±SD, n = 6) plasma concentration-time profiles of tanshinone IIA, cryptotanshinone, dihydrotanshinone I, salvianolic acid A, rosmarinic acid, ferulic acid, and lithospermic acid in rat plasma after oral administration of crude danshen (black) and wine-processed Danshen extracts (red). CDS, crude Danshen; WDS, wine-processed Danshen.

Table 5. Pharmacokinetic parameters of tanshinone IIA, cryptotanshinone, dihydrotanshinone I, salvianolic acid A, rosmarinic acid, ferulic acid, and lithospermic acid in rats after oral administration of crude Danshen and wine-processed Danshen (n = 6, mean ± SD) [a].

	Analytes	Doses (mg/kg)	$AUC_{0\text{-}t}$ [b] (ng·h/mL)	$AUC_{0\text{-}\infty}$ (ng·h/mL)	C_{max} (ng/mL)	CL/F (L/h/kg)	T_{max} (h)	MRT (h)	$T_{1/2}$ (h)
CDS	Tanshinone IIA	0.283	15.80 ± 5.62	15.80 ± 5.62	**9.36 ± 3.17** *	21.23 ± 9.60	0.58 ± 0.19	1.70 ± 0.29	1.73 ± 0.44
	Cryptotanshinone	0.196	30.92 ± 11.95	30.97 ± 12.01	18.48 ± 5.91	7.36 ± 2.81	0.46 ± 0.09	1.97 ± 0.54	2.82 ± 1.56
	Dihydrotanshinone I	0.64	**446.17 ± 94.44** *	**449.09 ± 92.53** *	**173.67 ± 31.12** *	**1.49 ± 0.28** *	0.83 ± 0.24	2.73 ± 1.39	3.34 ± 3.58
	Salvianolic acid A	2.17	343.16 ± 57.06	354.79 ± 56.29	87.92 ± 13.82	**6.50 ± 1.10** *	0.50 ± 0.00	5.84 ± 0.93	9.12 ± 2.48
	Rosmarinic acid	5.05	57.63 ± 12.38	61.58 ± 16.73	31.88 ± 6.69	93.35 ± 27.27	0.25 ± 0.00	4.05 ± 3.05	6.65 ± 6.91
	Ferulic acid	0.149	3.41 ± 1.35	3.41 ± 1.35	4.14 ± 1.00	49.94 ± 17.00	0.58 ± 0.19	0.69 ± 0.23	N.A.
	Lithospermic acid	12.40	1093.00 ± 179.05	1116.94 ± 183.96	249.00 ± 35.22	11.65 ± 1.88	0.67 ± 0.24	6.89 ± 1.14	7.33 ± 1.74
WDS	Tanshinone IIA	0.335	9.62 ± 2.85	9.60 ± 2.86	**3.94 ± 1.12** *	37.56 ± 9.28	0.67 ± 0.24	2.08 ± 0.80	2.06 ± 0.98
	Cryptotanshinone	0.231	23.78 ± 8.35	24.54 ± 9.77	11.59 ± 3.63	10.95 ± 3.71	0.54 ± 0.22	2.51 ± 1.60	3.29 ± 3.47
	Dihydrotanshinone I	0.723	**300.84 ± 78.25** *	**304.75 ± 79.61** *	**101.95 ± 15.35** *	**2.57 ± 0.70** *	0.67 ± 0.24	2.84 ± 0.97	4.30 ± 4.81
	Salvianolic acid A	1.66	340.75 ± 65.30	364.27 ± 70.23	73.57 ± 14.13	**17.27 ± 3.82** *	0.58 ± 0.19	8.03 ± 3.17	10.26 ± 4.30
	Rosmarinic acid	5.64	47.58 ± 12.78	47.62 ± 12.79	27.32 ± 6.88	132.83 ± 55.11	0.33 ± 0.12	1.91 ± 0.71	1.98 ± 1.36
	Ferulic acid	0.165	3.73 ± 1.63	3.73 ± 1.63	4.28 ± 1.25	55.9 ± 28.50	0.40 ± 0.12	0.62 ± 0.16	N.A.
	Lithospermic acid	10.1	1058.43 ± 256.90	1131.25 ± 227.92	238.67 ± 49.66	10.16 ± 2.58	0.58 ± 0.19	7.19 ± 1.61	10.51 ± 5.10

*, $p < 0.05$ compared with Crude group; [a], was assessed by the Kruskal–Wallis test; [b], significantly changed pharmacokinetic parameters in both groups are in bold; N.A., not applicable. CDS, crude Danshen; WDS, wine-processed Danshen.

4. Discussion and Conclusions

Danshen extract has been widely utilized in the treatment of cardiovascular-related diseases clinically [28]. Processing crude Danshen with wine has long been considered to enhance its biological activities; however, the mechanisms supported by adequate evidence are not clear yet. In the current study, we developed and optimized an HPLC-MS/MS method to simultaneously quantify seven bioactive components of Danshen in rat plasma for the first time. A model in which both positive and negative ions were detected simultaneously was built without losing the specificity and sensitivity (LLOQ = 0.5 ng/mL). Besides, the complete analysis takes a period as short as 8 min. With the sensitive method, a small quantity of biological sample (5 μL) at each time point was needed for the analysis, and no additional pre-column derivatization of the compounds was required. By employing a simple one-step protein precipitation approach for sample pretreatment and a relatively quick procedure for chromatographic separation (running time = 8 min), we were able to rapidly analyze the samples in a highly efficient manner. The establishment of the method is valuable for the quick quantification of various Danshen active components in biological samples, providing information about their pharmacokinetical behavior in vivo, explaining their efficiency model. Moreover, the study will direct the development of an analytical method for therapeutic monitoring the components, ensuring the safety and effectiveness of drug application.

Pharmacokinetic comparison between crude and wine-processed Danshen revealed that all the active ingredients in Danshen extract were absorbed fastly in vivo. This is consistent with their clinical application in the treatment of acute cardiovascular diseases. However, we observed significant decreases for C_{max} of tanshinone IIA and dihydrotanshinone I, and *AUC* for dihydrotanshinone I in wine-processed Danshen group. These results were completely contrary to our initial expectations: increased contents of tanshinone IIA and dihydrotanshinone I in wine-processed Danshen would result in higher plasma maximum concentration and absorption. By referring to previous literature, we speculate that the results might be attributed to the faster tissue distribution of the components in wine-processed Danshen group after initial absorption. As reported, processing of *Chuanxiong Rhizoma* with wine significantly decreased the $AUC_{0\text{-}t}$ values of its four active compounds, while increased their apparent volume of distribution (*Vd*), indicating wider tissue distribution [29]. Wine-processing also modified the distribution model of some flavonoids in *Radix scutellariae*: increased distribution in the rat upper-energizer tissues (lung and heart) and decreased distribution in the rat middle-and lower-energizer tissues (spleen, liver, and kidney) [30]. Thus, wine-processing might enhance the biological activity of Danshen by accelerating its distribution into target tissues. Detailed mechanisms about drug distribution into different tissues still need to be further investigated in subsequent studies.

Regarding the four hydrophilic acids, no significant differences between the pharmacokinetic parameters were found between the crude and wine-processed Danshen. A second peak was observed in the plasma-concentration profile of ferulic acid in the WDS group. We then did literature searching and found that the bimodal phenomenon was widely observed in pharmacokinetic profiles of constituents from traditional herbal medicine, which was probably due to distribution, reabsorption, and enterohepatic circulation [31,32]. Therefore, we guess that the second peak for ferulic acid in WDS might also be attributed to the above reasons. Further study is required to demonstrate the underlying mechanism. Our findings indicate that wine-processing of Danshen may modify the pharmacokinetic properties of some active components. However, the modification is limited, which might restrict its application in clinics. Since processing traditional herbal medicines is time-consuming and sometimes expensive, comparative study of the in vivo efficiency and pharmacokinetics between crude and processed medicines is rather necessary and meaningful. Our finding will provide useful suggestions for the doctors about the reasonable application of crude or processed Danshen.

Author Contributions: Conceptualization, S.W. and H.L.; methodology, Y.Z. and W.C.; software, Y.Z. and N.W.; validation, Y.Z., W.C., H.L. and S.W.; formal analysis, Y.Z., N.W. and W.C.; investigation, Y.Z., W.K., N.W. and W.C.; resources, Y.Z. and W.C.; data curation, Y.Z., W.K., N.W., J.S. and W.C.; writing—original draft preparation, Y.Z.; writing—review and editing, Y.Z. and J.S.; visualization, Y.Z. and W.C.; supervision, S.W. and H.L.; project administration, S.W. and H.L.; funding acquisition, X.L., S.W. and H.L. All authors have read and agreed to the published version of the manuscript.

Funding: This research was funded by Key Research and Development Project of Shandong Province (2017GSF218049), Young Scholars Program of Shandong University (2018WLJH93), and National Natural Science Foundation of Shandong Province (ZR2020MH374).

Institutional Review Board Statement: The experimental protocol was approved by the University Ethics Committee and conformed to the "Principles of Laboratory Animal Care" (NIH publication no. 85-23, revised 1985).

Informed Consent Statement: Not Applicable.

Data Availability Statement: All Data is contained within the article.

Acknowledgments: The authors gratefully acknowledge Xu Lingchuan for Danshen authentication.

Conflicts of Interest: The authors declare no conflict of interest.

References

1. Jung, I.; Kim, H.; Moon, S.; Lee, H.; Kim, B. Overview of salvia miltiorrhiza as a potential therapeutic agent for various diseases: An update on efficacy and mechanisms of action. *Antioxidants* **2020**, *9*, 857. [CrossRef]
2. Fang, Z.; Zhang, M.; Liu, J.; Zhao, X.; Zhang, Y.; Fang, L. Tanshinone IIA: A Review of its Anticancer Effects. *Front. Pharmacol.* **2021**, *11*, 2189. [CrossRef]
3. Li, Z.; Xu, S.; Liu, P. Salvia miltiorrhizaBurge (Danshen): A golden herbal medicine in cardiovascular therapeutics. *Acta Pharmacol. Sin.* **2018**, *39*, 802–824. [CrossRef] [PubMed]
4. Cao, M.; Liu, Y.; Jiang, W.; Meng, X.; Zhang, W.; Chen, W.; Peng, D.; Xing, S. UPLC/MS-based untargeted metabolomics reveals the changes of metabolites profile of Salvia miltiorrhiza bunge during Sweating processing. *Sci. Rep.* **2020**, *10*, 1–10. [CrossRef] [PubMed]
5. Wang, X.; Morris-Natschke, S.; Lee, K. New developments in the chemistry and biology of the bioactive constituents of Tanshen. *Med. Res. Rev.* **2007**, *27*, 133–148. [CrossRef] [PubMed]
6. Cai, N.; Chen, J.; Bi, D.; Gu, L.; Yao, L.; Li, X.; Li, H.; Xu, H.; Hu, Z.; Liu, Q.; et al. Specific degradation of endogenous Tau protein and inhibition of Tau fibrillation by tanshinone IIA through the ubiquitin-proteasome pathway. *J. Agric. Food Chem.* **2020**, *68*, 2054–2062. [CrossRef]
7. Ren, B.; Liu, Y.; Zhang, Y.; Zhang, M.; Sun, Y.; Liang, G.; Xu, J.; Zheng, J. Tanshinones inhibit hIAPP aggregation, disaggregate preformed hIAPP fibrils, and protect cultured cells. *J. Mater. Chem. B* **2017**, *6*, 56–67. [CrossRef] [PubMed]
8. Zhao, Y.; Du, X.; Duan, Y.; Pan, X.; Sun, Y.; You, T.; Han, L.; Jin, Z.; Shang, W.; Yu, J.; et al. High-throughput screening identifies established drugs as SARS-CoV-2 PLpro inhibitors. *Protein Cell* **2021**, 1–12. [CrossRef]
9. Li, X.; Du, F.; Jia, W.; Olaleye, O.; Xu, F.; Wang, F.; Li, L. Simultaneous determination of eight Danshen polyphenols in rat plasma and its application to a comparative pharmacokinetic study of DanHong injection and Danshen injection. *J. Sep. Sci.* **2017**, *40*, 1470–1481. [CrossRef]
10. Liu, X.; Jin, M.; Zhang, M.; Li, T.; Sun, S.; Zhang, J.; Dai, J.; Wang, Y. The application of combined 1 H NMR-based metabolomics and transcriptomics techniques to explore phenolic acid biosynthesis in Salvia miltiorrhiza Bunge. *J. Pharm. Biomed. Anal.* **2019**, *172*, 126–138. [CrossRef]
11. Zhang, C.; Pan, Y.; Cai, R.; Guo, S.; Zhang, X.; Xue, Y.; Wang, J.; Huang, J.; Wang, J.; Gu, Y.; et al. Salvianolic acid A increases the accumulation of doxorubicin in brain tumors through Caveolae endocytosis: The regulatory mechanism of BTB permeability induced by Salvianolic acid A. *Neuropharmacology* **2020**, *167*, 107980. [CrossRef] [PubMed]
12. Yang, Z.; Chen, Y.; Yan, Z.; Xu, T.; Wu, X.; Pi, A.; Liu, Q.; Chai, H.; Li, S.; Dou, X. Inhibition of TLR4/MAPKs pathway contributes to the protection of salvianolic acid A against lipotoxicity-induced myocardial damage in cardiomyocytes and obese mice. *Front. Pharmacol.* **2021**, *12*, 76. [CrossRef]
13. Chan, K.; Ho, W. Shing Anti-oxidative and hepatoprotective effects of lithospermic acid against carbon tetrachloride-induced liver oxidative damage in vitro and in vivo. *Oncol. Rep.* **2015**, *34*, 673–680. [CrossRef] [PubMed]
14. Singh, Y.; Rai, H.; Singh, G.; Singh, G.; Mishra, S.; Kumar, S.; Srikrishna, S.; Modi, G. A review on ferulic acid and analogs based scaffolds for the management of Alzheimer's disease. *Eur. J. Med. Chem.* **2021**, *215*, 113278. [CrossRef] [PubMed]
15. Li, M.; Cui, M.; Kenechukwu, N.; Gu, Y.; Chen, Y.; Zhong, S.; Gao, Y.; Cao, X.; Wang, L.; Liu, F.; et al. Rosmarinic acid ameliorates hypoxia/ischemia induced cognitive deficits and promotes remyelination. *Neural Regen. Res.* **2020**, *15*, 894–902. [CrossRef]

16. Ma, W.; Peng, Y.; Wang, W.; Bian, Q.; Wang, N.; Lee, D.Y.W.; Dai, R. Pharmacokinetic comparison of five tanshinones in normal and arthritic rats after oral administration of Huo Luo Xiao Ling Dan or its single herb extract by UPLC-MS/MS. *Biomed. Chromatogr.* **2016**, *30*, 1573–1581. [CrossRef]
17. Lu, P.; Xing, Y.; Xue, Z.; Ma, Z.; Zhang, B.; Peng, H.; Zhou, Q.; Liu, H.; Liu, Z.; Li, J. Pharmacokinetics of salvianolic acid B, rosmarinic acid and Danshensu in rat after pulmonary administration of Salvia miltiorrhiza polyphenolic acid solution. *Biomed. Chromatogr.* **2019**, *33*, e4561. [CrossRef]
18. Xie, X.; Miao, J.; Sun, W.; Huang, J.; Li, D.; Li, S.; Tong, L.; Sun, G. Simultaneous determination and pharmacokinetic study of four phenolic acids in rat plasma using UFLC–MS/MS after intravenous administration of salvianolic acid for injection. *J. Pharm. Biomed. Anal.* **2017**, *134*, 53–59. [CrossRef]
19. Lin, H.; Lin, T.; Chien, H.; Juang, Y.; Chen, C.; Wang, C.; Lai, C. A rapid, simple, and high-throughput UPLC-MS/MS method for simultaneous determination of bioactive constituents in Salvia miltiorrhiza with positive/negative ionization switching. *J. Pharm. Biomed. Anal.* **2018**, *161*, 94–100. [CrossRef]
20. Su, T.; Yu, H.; Kwan, H.; Ma, X.; Cao, H.; Cheng, C.; Leung, A.; Chan, C.; Li, W.; Cao, H.; et al. Comparisons of the chemical profiles, cytotoxicities and anti-inflammatory effects of raw and rice wine-processed Herba Siegesbeckiae. *J. Ethnopharmacol.* **2014**, *156*, 365–369. [CrossRef]
21. Sun, T.; Zhang, H.; Li, Y.; Liu, Y.; Dai, W.; Fang, J.; Cao, C.; Die, Y.; Liu, Q.; Wang, C.; et al. Physicochemical properties and immunological activities of polysaccharides from both crude and wine-processed Polygonatum sibiricum. *Int. J. Biol. Macromol.* **2020**, *143*, 255–264. [CrossRef] [PubMed]
22. Huang, Z.; Jiang, M.; Yi, Y.; Zeng, R.; Huang, Y.; Wu, P. Effects of Processed Radix Salviae Miltiorrhizae and Radix et Rhizoma Rhei with Wine on Functions of Blood Platelet and Anticoagulation of Rat. *Chin. Tradit. Pat. Med.* **2001**, *23*, 341–342. [CrossRef]
23. Li, C.; Zhao, L.; Yang, Y.; Kang, W. Antimicrobial activity of Salvia miltiorrhiza and different processed products. *Chin. Tradit. Pat. Med.* **2011**, *33*, 1948–1951. [CrossRef]
24. Cui, W.; Li, H.; Zhang, X.; Song, M.; Diao, J.; Zhang, D.; Wang, X. Analysis of five qualitative change compounds before and after wine processing of Salvia miltiorrhiza by UPLC-QE/MS. *Chin. Tradit. Pat. Med.* **2019**, *41*, 844–849. [CrossRef]
25. FDA Guidance for Industry Bioanalytical Method Validation Guidance for Industry Bioanalytical Method Validation. 2018, 1–44. Available online: https://www.fda.gov/regulatory-information/search-fda-guidance-documents/bioanalytical-method-validation-guidance-industry..
26. Zhang, D.; Sun, L.; Li, H.; Cui, Y.; Liu, S.; Wu, P.; Zhao, D.; Zhao, P.; Zhang, X. Pharmacokinetic comparison of nine bioactive components in rat plasma following oral administration of raw and wine-processed Ligustri Lucidi Fructus by ultra-high-performance liquid chromatography coupled with triple quadrupole mass spectrometry. *J. Sep. Sci.* **2020**, *43*, 3995–4005. [CrossRef] [PubMed]
27. Shi, B.; Li, Q.; Feng, Y.; Dai, X.; Zhao, R.; Zhao, Y.; Jia, P.; Wang, S.; Yu, J.; Liao, S.; et al. Pharmacokinetics of 13 active components in a rat model of middle cerebral artery occlusion after intravenous injection of Radix Salviae miltiorrhizae-Lignum dalbergiae odoriferae prescription. *J. Sep. Sci.* **2020**, *43*, 531–546. [CrossRef]
28. Hung, Y.; Wang, P.; Lin, T.; Yang, P.; You, J.; Pan, T. Functional redox proteomics reveal that salvia miltiorrhiza aqueous extract alleviates adriamycin-induced cardiomyopathy via inhibiting ROS-dependent apoptosis. *Oxid. Med. Cell. Longev.* **2020**, *2020*, 5136934. [CrossRef]
29. Ning, Y.; Pei, K.; Cao, G.; Cai, H.; Liu, X.; Cao, L.; Zhang, S.; Cai, B. Comparative study on pharmacokinetics of four active compounds in rat plasma after oral administration of raw and wine processed chuanxiong rhizoma. *Molecules* **2020**, *25*, 93. [CrossRef] [PubMed]
30. Huang, P.; Tan, S.; Zhang, Y.; Li, J.; Chai, C.; Li, J.; Cai, B. The effects of wine-processing on ascending and descending: The distribution of flavonoids in rat tissues after oral administration of crude and wine-processed Radix scutellariae. *J. Ethnopharmacol.* **2014**, *155*, 649–664. [CrossRef] [PubMed]
31. Du, Y.; He, B.; Li, Q.; He, J.; Wang, D.; Bi, K. Simultaneous determination of multiple active components in rat plasma using ultra-fast liquid chromatography with tandem mass spectrometry and application to a comparative pharmacokinetic study after oral administration of Suan-Zao-Ren decoction and Suan-Zao-Ren granule. *J. Sep. Sci.* **2017**, *40*, 2097–2106. [CrossRef] [PubMed]
32. Zhang, H.; Hu, X.; Qiao, M.; Li, Y.; Cao, S.; Ding, L.; Feng, X.; Kang, N.; Zhang, D.; Qiu, F. Simultaneous determination of five isoflavones in rat plasma by LC-MS/MS: Comparative pharmacokinetic characteristics of Puerariae lobatae radix in normal and type 2 diabetic rats. *J. Sep. Sci.* **2019**, *42*, 2592–2601. [CrossRef] [PubMed]

separations

MDPI

Article

Identification of Abnormal Proteins in Plasma from Gout Patients by LC-MS/MS

Lijin Shen [1], Hanyang Dong [2,*], Zhenchang Guo [2], Guijin Zhai [2] and Kai Zhang [2,*]

1 NHC Key Laboratory of Hormones and Development, Tianjin Key Laboratory of Metabolic Diseases, Chu Hsien-I Memorial Hospital & Tianjin Institute of Endocrinology, Tianjin Medical University, Tianjin 300134, China; shenlijin2018@163.com

2 The Province and Ministry Co-Sponsored Collaborative Innovation Center for Medical Epigenetics, Tianjin Key Laboratory of Medical Epigenetics, Key Laboratory of Immune Microenvironment and Disease (Ministry of Education), Tianjin Key Laboratory of Cellular Homeostasis and Diseases, Department of Biochemistry and Molecular Biology, Tianjin Medical University, Tianjin 300070, China; guozhenchang@tmu.edu.cn (Z.G.); zhaiguijin@tmu.edu.cn (G.Z.)

* Correspondence: donghanyang@tmu.edu.cn (H.D.); kzhang@tmu.edu.cn (K.Z.); Tel.: +86-22-8333-6833 (K.Z.)

Abstract: A high level of uric acid may cause hyperuricemia, which further develops into gout, eventually leading to chronic kidney disease. However, the pathogenic mechanism remains largely unknown. To investigate the cause and block the transformation of hyperuricemia to related diseases, it is important to discover the alterations in protein levels between gout patients and non-gout individuals. To date, human blood plasma is still the predominant matrices for clinical analysis. Due to the high abundance, the proteins of plasma samples have strong shielding effects on low abundance proteins, thus, the information on low abundance protein expression is always masked, while the low abundance proteins of human plasma are often of great significance for the diagnosis and treatment of diseases. Therefore, it is very important to separate and analyze the plasma proteins. High-performance liquid chromatography (LC) tandem mass spectrometry (MS)-based proteomics has been developed as a powerful tool to investigate changes in the human plasma proteome. Here, we used LC-MS/MS to detect the differential proteins in the plasmas from simple gout patients, gout with kidney damage patients, and non-gout individuals. We identified 32 obviously differential proteins between non-gout and gout subjects and 10 differential proteins between simple gout and gout with kidney damage patients. These differential proteins were further analyzed to characterize their localization and functions. Additionally, the correlation analysis showed multiple relationships between the abnormal plasma proteins and clinical biochemical indexes, particularly for the immune-inflammatory response proteins. Furthermore, inflammation factors gelsolin (GSN) were confirmed. Our results offer a view of plasma proteins for studying biomarkers of gout patients.

Keywords: proteomics; high-performance liquid chromatography; mass spectrometry; gout; uric acid

Citation: Shen, L.; Dong, H.; Guo, Z.; Zhai, G.; Zhang, K. Identification of Abnormal Proteins in Plasma from Gout Patients by LC-MS/MS. *Separations* **2021**, *8*, 85. https://doi.org/10.3390/separations8060085

Academic Editor: Victoria Samanidou

Received: 11 May 2021
Accepted: 6 June 2021
Published: 16 June 2021

Publisher's Note: MDPI stays neutral with regard to jurisdictional claims in published maps and institutional affiliations.

1. Introduction

Uric acid, a weak diprotic acid, is the terminal product of purine metabolism in humans [1]. Hyperuricemia, a high level of serum uric acid, is known to cause significant health problems associated with urate crystal deposition. Emerging evidence indicates that hyperuricemia is an independent risk factor for chronic kidney disease (CKD) [2,3]. Meanwhile, persistent hyperuricemia is an important factor in the development of gout, with about 5 to 12% of hyperuricemia developing into gout. It is often stated that the prevalence of hyperuricemia and gout has increased in recent years [4–6]. The prevalence of gout and hyperuricemia has increased in developed countries over the past two decades and research into the area has become progressively more active. According to a 2014 study, the adjusted prevalence of hyperuricemia among Chinese adults in 2009–2010 was about 8.4–13.3% [7]. The prevalence of gout ranged from 1 to 4% worldwide and incidence ranged from about 0.1 to 0.3% [8], while the prevalence rate of gout in China is 1–3%. Many epidemiological studies have shown that hyperuricemia and gout are associated with the development of hypertension, cardiovascular disease, chronic kidney disease, and so on [2]. Therefore, it is particularly important to find the cause and block the transformation of hyperuricemia into related diseases.

As the basic executor of life activities, proteins drive various physiological or pathological processes. As one of the largest subgroups of the human proteome, plasma proteomics has attracted more and more attention. To date, human blood plasma is still the predominant matrices for clinical analysis as they are easily accessible and clearly reflect an individual's metabolism, better than serum, thus, it is crucial to profile the differential plasma proteins between gout patients and non-gout individuals to understand the molecular mechanism of gout occurrence and development. Due to their high abundance, proteins in plasma samples have strong shielding effects on low abundance proteins, thus, the information on low abundance protein expression is always masked. However, the low abundance proteins of human plasma are often of great significance for the diagnosis and treatment of diseases. Therefore, it is very important to separate and enrich the low abundance proteins and remove the high abundance proteins from the plasma. HPLC-MS has been developed as a powerful tool to investigate changes in the human plasma proteome because this holistic approach can, in principle, yield specific and quantitative information on all proteins in an unbiased way [9]. This proteomic workflow allows the streamlined analysis of hundreds of plasma proteins from a single drop of blood.

The profile of plasma proteins can provide quantitative information on the majority of the classical, functional plasma proteins, and thus we speculated that the metabolic status of individuals during gout would be reflected by their plasma proteomes. Herein we selected 16 plasma samples of gout patients (8 cases of gout alone and 8 cases of gout with renal injury) and 8 cases of non-gout individuals for proteomic analysis. The biological significance of abnormal proteins was further analyzed by combining them with clinical syndromes and biochemical indicators. This allowed us to analyze the global changes related to gout processes in the plasma proteome.

2. Materials and Methods

2.1. Study Design

Sixteen patients with primary gout (non-acute stage) were collected in 2017 from the Gout Department of Tianjin Medical University Metabolic Diseases Hospital, including 8 male gout patients (G) and 8 male gout patients combined with renal injury (GN). At the same time, 8 samples from male health examinations in the center acted as a normal control group (NC); all of them had no history of gout, hyperuricemia, and kidney disease.

The diagnostic criteria for gout are based on the 2015 ACR/EULAR gout classification criteria. Participation in the investigation was voluntary.

2.2. Methods

2.2.1. The Plasma Sample Collection and Highly Abundant Protein Depletion

The fasted overnight blood samples were collected from 8 non-gout subjects as the normal control group (NC), 8 male gout patients alone (G), and 8 male gout patients combined with renal injury (GN). Then the collected blood samples were centrifuged for 10 min at 4 °C at 2500 rpm, and the upper plasma samples were stored at −80 °C.

For LC-MS/MS analysis, the highly abundant proteins of plasma were depleted using an Albumin/IgG removal column (Merck, Germany) according to the manufacturer's protocol. Briefly, 35 μL plasma mixed with 315 μL Binding Buffer was loaded into the activated columns, then we collected the effluent from the column as the plasma sample removed the highly abundant proteins.

2.2.2. Plasma Protein Sample Preparation

The plasma protein concentration was determined by the BCA method (Thermo Fisher Scientific, Waltham, MA, USA), Each plasma sample (about 100 μg) was dried and then the proteins were dissolved in 30 μL 50 mM NH_4HCO_3. After 20 min of denaturing on ice, the samples were diluted by 150 μL 50 mM NH_4HCO_3. Next, tryptic digestion (Promega, Madison, WI, USA) with 1: 50 (*w:w*) enzyme to substrate ratio was carried out at 37 °C, overnight. We then added 0.9 μL 1M dithiothreitol solution into the resulting peptides and the mixture was incubated at 56 °C for 1 h. After cooling to room temperature, the peptides were alkylated by 5.4 μL 500 mM iodoacetamide in a dark room for 45 min. After 5.4 μL 500 mM cysteine was added into the peptide mixture for 30 min, the trypsin with 1:100 (*w:w*) enzyme to substrate ratio was added and incubated for 4 h. The digestion was terminated by adding 10 μL 10% (*v/v*) trifluoroacetic acid (TFA) and the portion of resulting peptides (about 2 μg) were dried, and cleaned with U-C18 ZipTips (Millipore Corp.) according to the manufacturer's instructions. Briefly, the column was first activated with 100% acetonitrile, transited with 50% acetonitrile, and balanced with 0.1% TFA. Then, the sample was loaded and the column was desalted with 0.1% TFA. Next the column was eluted with 50% acetonitrile, and finally, the elution was dried for nano-HPLC/mass spectrometric analysis.

2.2.3. LC-MS/MS Analysis

Analysis was performed in triplicate. Each sample of peptides was reconstituted in 7 μL of HPLC buffer A (0.1% (*v/v*) formic acid in water), and 5 μL was injected into a Nano-LC system (EASY-nLC 1000, Thermo Fisher Scientific, Waltham, MA, USA). Each sample was separated by a C18 column (50 μm inner-diameter × 15 cm, 2 μm C18) with a 120 min HPLC-gradient (linear gradient from 2 to 35% HPLC buffer B (0.1% formic acid in acetonitrile) for 110 min, and then to 90% buffer B for 10 min). The separation was carried out at 19 °C and the HPLC elution was electrosprayed directly into an Orbitrap Q-Exactive mass spectrometer (Thermo Fisher Scientific, Waltham, MA, USA). The source spray voltage was set at 1.8 kV and the capillary temperature was set at 320 °C. The mass spectrometric analysis was carried out in a data-dependent mode with an automatic switch between a full MS scan and an MS/MS scan in the orbitrap. For the full MS survey scan, the automatic gain control (AGC) target was 3×10^6 scan range was from 350 to 1750 with a resolution of 70,000. The 15 most intense peaks with charge state 2 and above were selected for fragmentation by higher-energy collision dissociation (HCD) with a normalized collision energy of 27%. The MS2 spectra were acquired with 17,500 resolution. The exclusion duration for the data-dependant scan was 30 s, the repeat count was 2.

2.2.4. Database Searching and Bioinformatics Analysis

Data analysis: The resulting MS/MS data were searched against UniProt database using Maxquant software (V1.5.5) with an overall false discovery rate (FDR) for peptides of less than 1%. Peptide sequences were searched using trypsin specificity and allowing a maximum of two missed cleavages. Label-free protein quantification (LFQ).

Bioinformatics analysis: Categorical annotation was supplied in the form of gene ontology (GO) biological process (BP) and cellular component (CC). The distribution of different proteins in metabolic pathways by Kyoto Encyclopedia of Genes and Genomes (KEGG) Pathway. Protein–protein interactions were performed by STRING.

2.2.5. Validation of Identified Proteins by ELISA

Plasma samples from the normal control group, gout alone, and gout with renal injury were tested using Human GS (Gelsolin) ELISA kit (Elabscience Biotechnology Co., Ltd., Wuhan, China). The ELISA protocols provided by the manufacturers were strictly followed. For the ELISA, the plasma samples (diluted 1:4) were placed in wells coated with a GS antigen and incubated for 1.5 h at 37 °C. After a wash to eliminate the non-associated components, an anti-GS/peroxidase conjugate was added, and the plates were incubated for 1 h at 37 °C. Following the addition of the stop solution, the optical density was measured at 450 nm using a multimode reader (Thermo, USA).

2.2.6. Statistical Analysis

All data were expressed as mean ± S.D. Statistical analyses were performed using t-test of two independent samples with the measurement data. And $p < 0.05$ was considered statistically significant.

3. Results

3.1. Profile of Plasma Proteomes in Gout Study

To obtain the profile of proteins in the gout study, we took plasma samples from 16 patients with gout (including 8 patients with gout alone and 8 patients with gout accompanied by renal injury) and from 8 normal people, and thus the formed three groups of samples. The basic clinical data from gout patients was collected to analyze the relationship between differential proteins and gout pathology, as shown in Tables 1 and 2. To achieve the depth analysis of proteomes, we performed the deletion of the abundant plasma proteins by using Albumin/IgG removal column separation. The purified proteins were subjected to in-solution tryptic digestion, subsequently resulting peptides were separated and analyzed by LC-MS/MS. The raw data was further analyzed using Maxquant software and bioinformatics tools (Figure 1).

Table 1. Comparison of clinical data between the control group and gout group ($\overline{x} \pm$ s).

Clinical Data	Normal Control	Gout	t	p
Numbers	8	16	-	-
Age	57.63 ± 1.19	55.88 ± 6.83	0.995	0.334
BMI (kg/m^2)	24.74 ± 2.50	28.08 ± 4.86	−1.815	0.083
SBP (mmHg)	121.88 ± 14.05	135.31 ± 21.01	−1.627	0.118
DBP (mmHg)	71.13 ± 14.59	87.31 ± 14.02 *	−2.631	0.015
SUA (umol/L)	328.50 ± 47.91	497.98 ± 126.85 **	−4.714	0.000
Scr (umol/L)	88.00 ± 5.83	97.94 ± 22.45	−1.663	0.113
BUN (mmol/L)	6.33 ± 1.27	5.88 ± 1.48	0.728	0.474
eGFR (ml/min)	100.73 ± 7.79	95.65 ± 23.49	0.784	0.442
TG (mmol/L)	1.02 ± 0.38	2.83 ± 1.65 **	−4.178	0.001
CHO (mmol/L)	4.50 ± 0.65	5.55 ± 0.71 **	−3.510	0.002
ALT (IU/L)	18.38 ± 3.70	27.85 ± 7.92 **	−3.188	0.004
FPG (mmol/L)	5.10 ± 0.30	5.29 ± 0.86	−0.771	0.450

Compared with the normal control group, * $p < 0.05$, ** $p < 0.01$. The parameters include: BMI (body mass index), SBP (systolic blood pressure), DBP (diatolic blood pressure), SUA (serum uric acid), Scr (serum creatintine), BUN (blood urea nitrogen), eGFR (estimated glomerular filtration rate), TG (triglyceride), CHO (cholesterol), ALT (alanine aminotransferase), and FPG (fasting plasma glucose).

Table 2. Comparison of clinical data between gout alone and gout with renal injury ($\overline{x} \pm$ s).

Clinical Data	G	GN	t	p
Numbers	8	8	-	-
Age	57.50 ± 6.72	54.25 ± 6.98	0.948	0.359
Course	14.38 ± 7.80	10.00 ± 6.32	1.232	0.238
BMI (kg/m^2)	25.98 ± 2.53	30.19 ± 5.84	−1.872	0.082
SBP (mmHg)	131.88 ± 27.50	138.75 ± 12.75	−0.641	0.536
DBP (mmHg)	83.13 ± 17.51	91.50 ± 8.65	−1.213	0.245
SUA (umol/L)	491.4 ± 151.1	504.56 ± 107.47	−0.201	0.844
Scr (umol/L)	78.76 ± 8.22	117.13 ± 13.09 **	−7.020	0.000
BUN (mmol/L)	5.07 ± 0.99	6.69 ± 1.47 *	−2.58	0.022
eGFR(ml/min)	116.30 ± 11.23	74.99 ± 9.01 **	8.116	0.000

Table 2. *Cont.*

Clinical Data	G	GN	*t*	*p*
TG (mmol/L)	3.46 ± 1.93	2.20 ± 1.10	1.593	0.133
CHO (mmol/L)	5.40 ± 0.57	5.71 ± 0.84	−0.851	0.409
ALT (IU/L)	24.99 ± 4.40	30.71 ± 9.81	−1.506	0.154
AST (IU/L)	24.49 ± 5.97	25.34 ± 7.90	−0.243	0.812
FPG (mmol/L)	5.04 ± 0.75	5.53 ± 0.95	−1.154	0.268
HOMA-IR	2.30 ± 0.49	3.84 ± 1.38 *	−2.955	0.016

Compared with gout alone (G), * $p < 0.05$, ** $p < 0.01$. The parameters include: BMI (body mass index), SBP (systolic blood pressure), DBP (diatolic blood pressure), SUA (serum uric acid), Scr (serum creatintine), BUN (blood urea nitrogen), eGFR (estimated glomerular filtration rate), TG (triglyceride), CHO (cholesterol), ALT (alanine aminotransferase), AST (aspartate aminotransferase), FPG (fasting plasma glucose), HOMA-IR (homeostasis model assessment-insulin resistance).

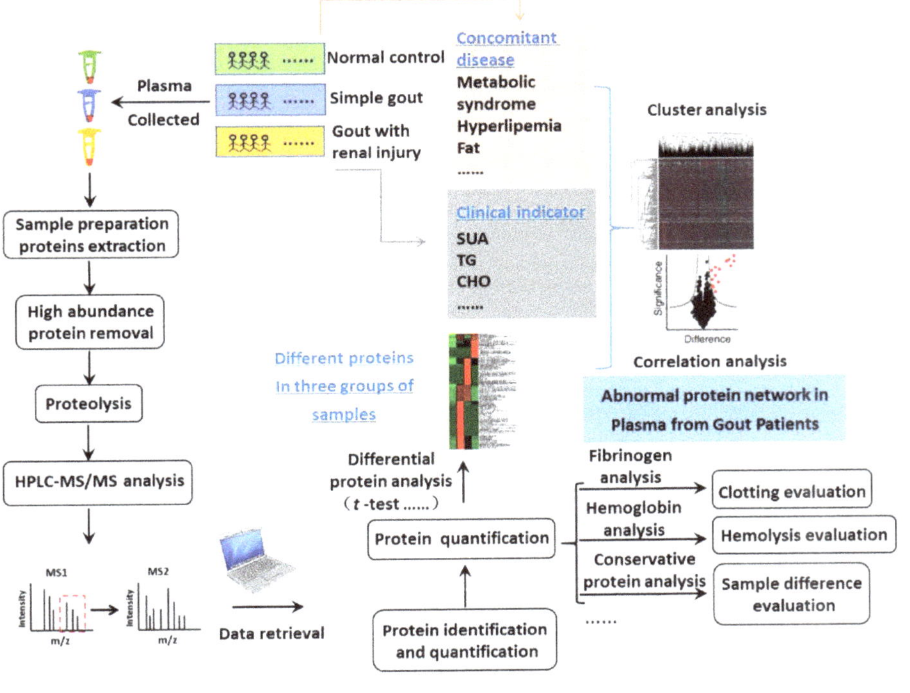

Figure 1. Study design and workflow. The study cohort consisted of 24 individuals. Triplicates of 24 plasma samples resulted in 72 plasma proteomes. The LC-MS/MS data, which we analyzed by MaxQuant.

To obtain accurate quantification, we analyzed triplicates of 24 plasma samples. Finally, 314 proteins were identified in the normal control group, 303 proteins were identified in the gout group, 283 proteins were identified in the gout with kidney injury group; 369 proteins (subtracting contaminants such as keratins) were identified in three groups. These proteins were mainly located in blood particles, adhesive plaques, extracellular matrix, etc., and participated in innate immune response, platelet degranulation, protein hydrolysis, and the classical activation pathway of complement (Figure 2).

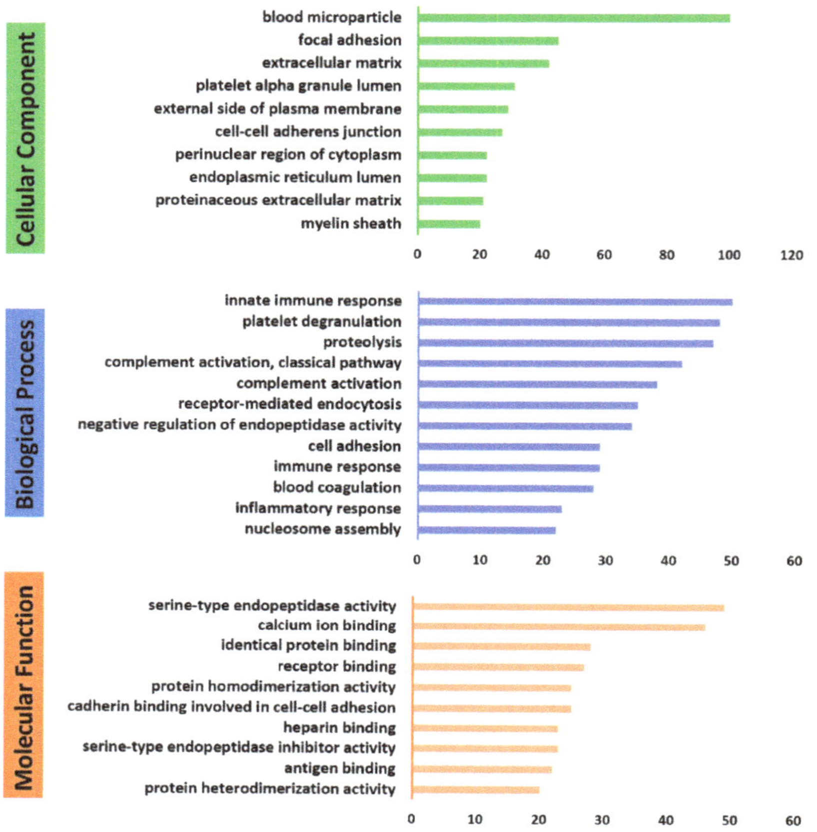

Figure 2. Among the 24 plasma samples, analysis of the biological process, molecular function, and cellular component of plasma differential proteins.

On the basis of quantitative proteomics, our measurements contained much important information about the samples. For instance, we use fibrinogen, which has a coagulation function, as the basis for the evaluation of coagulation. Of the 24 plasma samples we measured, there was no significant difference in fibrinogen FGA, FGB, and FGG in 24 individuals (Figure 3A). To see if the sample had hemolytic events, we used high abundance hemoglobin as the evaluation index. There was no significant increase in the expression of hemoglobin HBA1, HBB, HBD, and CA1 in the 24 individuals, as shown in Figure 3B. These results suggest that there was no obvious coagulation and hemolysis in these samples and excellent sample handling procedures throughout the study.

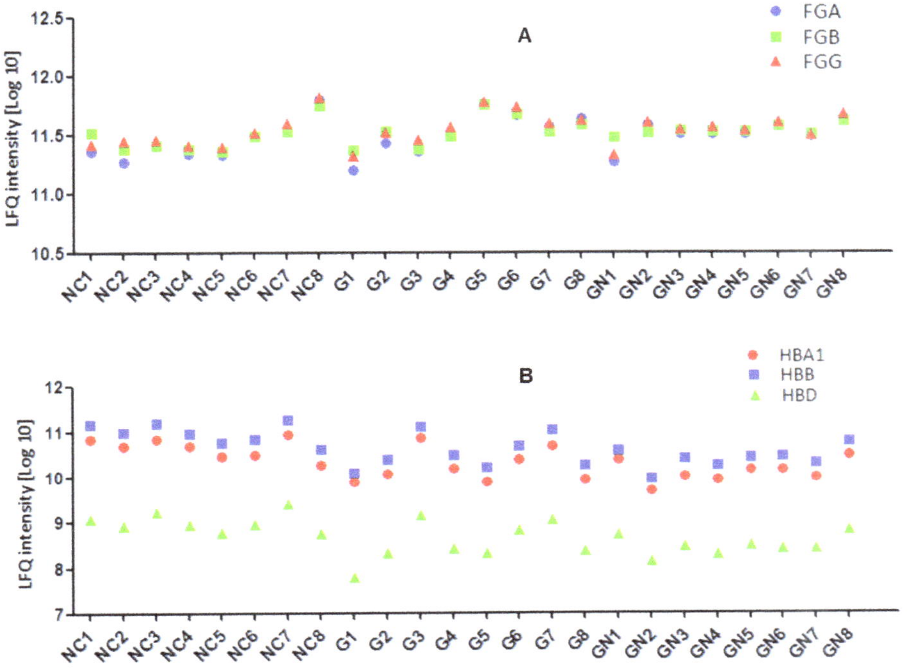

Figure 3. Among the 24 plasma samples, we measured the evaluation of plasma coagulation and hemolysis events. (**A**) Analysis of protein abundance of FGA, FGB, and FGG in 24 plasma samples. (**B**) Analysis of protein abundance of HBA1, HBB, and HBD in 24 plasma samples.

3.2. Plasma Protein Levels Are Individual-Specific

Overall, protein levels tended to vary considerably between participants; it is believed to one of the main problems in proteomics analysis. In our analysis, lipoprotein(a) (LPA), C-reactive (CRP), and pregnancy zone protein (PZP) were 10 times higher in some individuals compared to others (Figure 4A). A total of 83 proteins were detected in a single individual. The differences among some proteins are very low, such as hemoglobin binding protein (HPX), complement factor C6, cluster protein (CLU), and L-selectin (SELL), which showed that biology attributes of samples were highly conserved (Figure 4B).

3.3. Comparison of Proteins in Plasma between Normal Control Group and Gout Group

To study the abnormal proteins in plasma from gout patients, we compared plasma proteins between the normal control group and gout patient group. Two independent sample *t*-test methods were used ($p < 0.05$), and thus 32 differential proteins were identified (Table 3). Compared with the control group, the gout patient group had a comprehensive effect on the blood plasma proteome profile, with 22 decreased and 10 increased protein levels (Figure 5A). According to their function, these proteins were clustered into five groups (Figure 5B). Group 1 comprised of glycolipid proteins and cellular activity proteins. They were mainly apolipoproteins (APOC4, APOD) and GAPDH, VTN, GSN, FN1, and TMSB4X. Group 2 comprised of the complement (C3, C4BPB, C5, C8B, C8A, and C8G). Group 3 comprised of enzymes (LTF, CA1, PPIAP, and GLYRP2). Group 4 comprised of oxidative damage and inflammatory markers (S100A8, S100A9, PRDX2, ORM2, GPX3, CRP, and ANXA1). Group 5 comprised of other functional proteins (ITIH4, IGFBP4,

CLEC3B, LBP, Ig, and LCP1). The interaction networks analysis further showed that the 32 differential proteins wre closely related (Figure 5C).

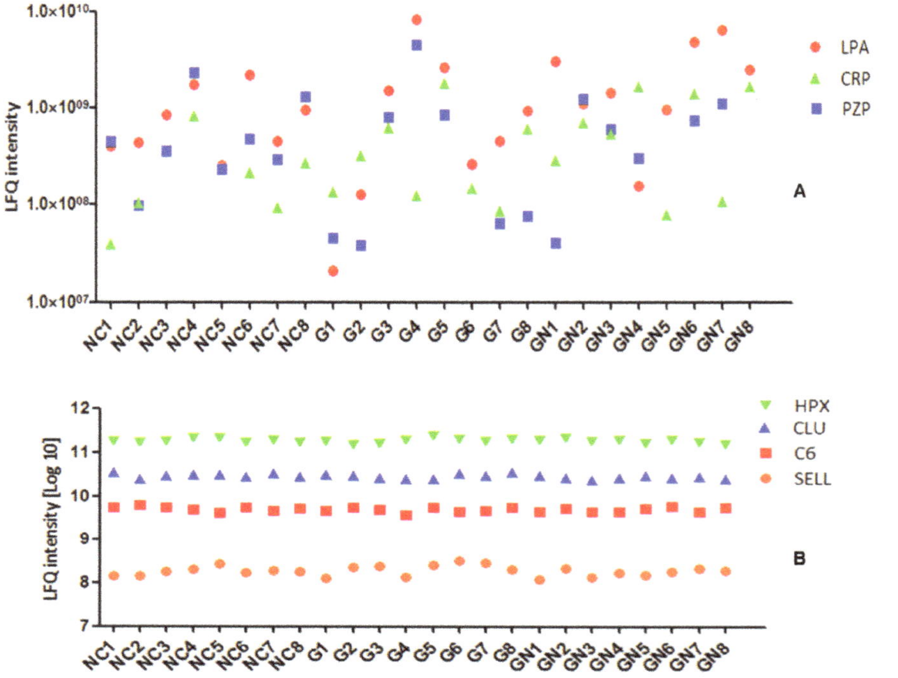

Figure 4. Individual differences in plasma protein levels among the 24 plasma samples. (**A**) Analysis of protein abundance of LPA, CRP, and PZP in 24 plasma samples. (**B**) Analysis of protein abundance of SELL, C6, CLU, and HPX in 24 plasma samples.

Table 3. Differential proteins in the normal control group and gout group.

Majority Protein IDs	Protein Names	Gene Names	*t*	*p*
P00915	Carbonic anhydrase 1	CA1	2.467	0.037
P01024	Complement C3	C3	−2.522	0.019
P01031	Complement C5	C5	2.556	0.018
P0DP04	Ig heavy chain V-III region DOB		−2.201	0.039
P02741	C-reactive protein	CRP	−2.407	0.025
P02751	Fibronectin	FN1	−2.111	0.046
P02788	Lactotransferrin	LTF	2.522	0.019
P04004	Vitronectin	VTN	−2.830	0.010
P04083	Annexin A1	ANXA1	2.403	0.047
P04196	Histidine-rich glycoprotein	HRG	2.298	0.031
P04406	Glyceraldehyde-3-phosphate dehydrogenase	GAPDH	5.462	0.000
P05090	Apolipoprotein D	APOD	2.860	0.009
P05109	Protein S100-A8	S100A8	3.241	0.014
P05452	Tetranectin	CLEC3B	2.174	0.041
P06396	Gelsolin	GSN	2.277	0.033
P06702	Protein S100-A9	S100A9	4.003	0.005
P07357	Complement component C8 alpha chain	C8A	2.080	0.049
P07358	Complement component C8 beta chain	C8B	3.924	0.001

Table 3. *Cont.*

Majority Protein IDs	Protein Names	Gene Names	*t*	*p*
P07360	Complement component C8 gamma chain	C8G	2.389	0.026
P13796	Plastin-2	LCP1	4.047	0.001
P18428	Lipopolysaccharide-binding protein	LBP	2.333	0.029
P19652	Alpha-1-acid glycoprotein 2	ORM2	−2.371	0.027
P20851	C4b-binding protein beta chain	C4BPB	2.263	0.034
P22352	Glutathione peroxidase 3	GPX3	2.568	0.018
P22692	Insulin-like growth factor-binding protein 4	IGFBP4	−2.183	0.040
P32119	Peroxiredoxin-2	PRDX2	3.061	0.015
P55056	Apolipoprotein C-IV	APOC4	−2.972	0.007
P62328	Thymosin beta-4	TMSB4X	2.204	0.038
P62937	Peptidyl-prolyl cis-trans isomerase A	PPIA	2.295	0.032
P80748	Ig lambda chain V-III region LOI		−3.514	0.002
Q14624	Inter-alpha-trypsin inhibitor heavy chain H4	ITIH4	3.851	0.001
Q96PD5	N-acetylmuramoyl-L-alanine amidase	PGLYRP2	−2.307	0.031

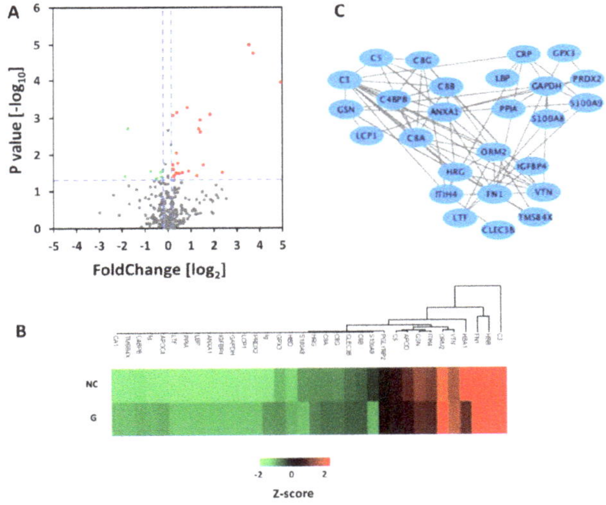

Figure 5. Comparison of differential proteins between the normal control group and gout group among the 24 plasma samples. (**A**) Volcanic diagram analysis of differential protein between the normal control group and gout group. (**B**) Cluster analysis of 32 differentially expressed proteins in plasma of patients with the normal control group and gout group. (**C**) Network correlation analysis of 32 differentially expressed proteins in plasma of patients with the normal control group and gout group.

3.4. Comparison of Proteins in Plasma between Gout and Gout with Renal Injury

To further explore the functional effect of plasma proteins on gout development, among the gout patients, we compared plasma proteins between patients with gout alone and patients with renal injury. Using two independent samples *t*-test method ($p < 0.05$), we found 10 differential proteins (Table 4). The results show that the expressions of plasma complement C4A, C4B, and SERPINF1 in patients with gout accompanied by renal injury are significantly increased than those of patients with gout alone.

Table 4. Differential proteins in simple gout group and gout with renal injury group.

Majority Protein IDs	Protein Names	Gene Names	*t*	*p*
A0A0B4J1U7	Ig	IGHV6-1	2.735	0.023
A0A0C4DH67	Ig	IGKV1-8	2.987	0.013
P01833	Polymeric immunoglobulin receptor	PIGR	2.465	0.039
P02656	Apolipoprotein C-III	APOC3	2.157	0.049
P05155	Plasma protease C1 inhibitor	SERPING1	−2.400	0.031
P05546	Heparin cofactor 2	SERPIND1	2.612	0.021
POCOL4	Complement C4-A	C4A	−3.091	0.008
POCOL5	Complement C4-B	C4B	−2.450	0.028
P36955	Pigment epithelium-derived factor	SERPINF1	−2.565	0.022
Q92954	Proteoglycan 4	PRG4	2.667	0.018

3.5. The Relationship between Plasma Protein and Clinical Biochemical Indexes

Next, the quantified plasma proteins were correlated with classical laboratory parameters, including serum uric acid (SUA), eGFR, Cr, BUN, FPG, BMI, CHO, TG, and ALT. Remarkably, of all proteins in our data, the five proteins most significantly correlating with SUA were inflammation factors Gelsolin (GSN), S100A8, S100A9, ORM2, and ANXA1. For each, the *p*-value was lower than 0.05 and Pearson correlation coefficients ranged from 0.4 to 0.6. GSN, S100A8, S100A9, and ANXA1 were negatively correlated with serum uric acid. Secondly, the proteins significantly related to CHO were CRP, GSN, S100A8, S100A9, ORM2, ANXA1, PRDX2, and GPX3. Pearson's correlation coefficient was 0.4–0.7. Thirdly, the proteins significantly related to TG were S100A8, S100A9, ORM2, PRDX2, C3, C8A, C8B, and FN1. Pearson's correlation coefficient was 0.3−0.6 (Figure 6A). These data showed the functional abnormal proteins of gout plasma and thus indicated potential markers of plasma for clinical analysis.

3.6. Validation of GSN in Individual Subjects

To validate the accuracy of our LC-MS/MS results, a key protein, GSN, was selected for further study by ELISA. Consistent with the results of the data in the Table 3 and Figure 6A, the protein level of GSN was significantly downregulated in the gout groups compared with the normal control group (Figure 6B).

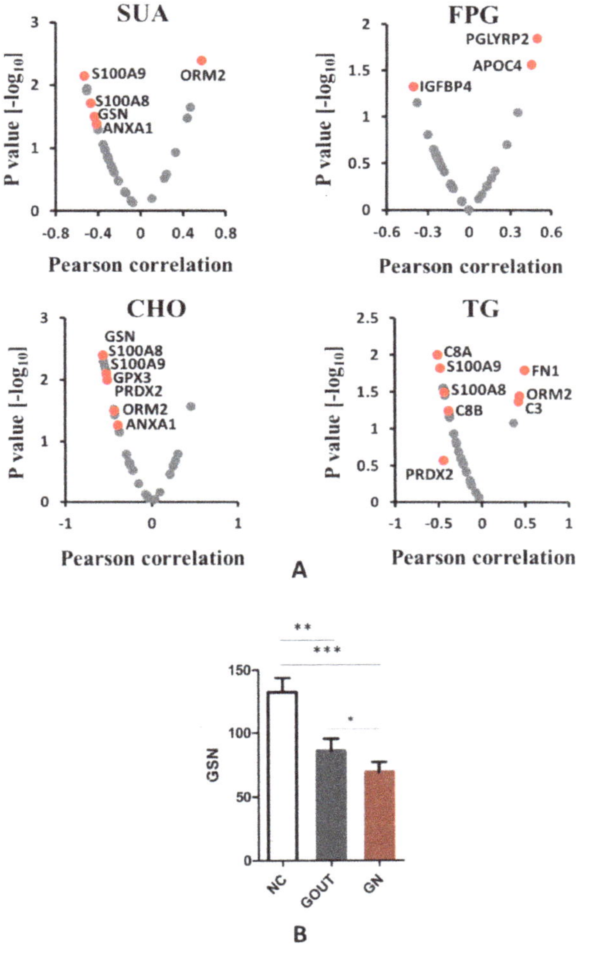

Figure 6. (**A**) Correlation between abnormal plasma proteins and clinical biochemical index in patients with gout. (**B**) ELISA confirmation of protein GSN between the normal control group and gout groups. Data were expressed as the mean using GraphPad Prism. * $p < 0.05$, ** $p < 0.01$, *** $p < 0.0001$.

4. Discussion

Clinical analysis of blood is the most widespread diagnostic procedure in medicine [10]. As one of the largest subgroups of the human proteome, plasma proteomics has attracted more and more attention. LC-MS/MS has become a powerful technique in biological research, which will make the study of proteomics in-depth.

There are many kinds of proteins in human plasma, which not only contain a lot of albumin and immunoglobulin, but also contain some low concentration proteins. Because of the existence of high abundance proteins (especially albumin) in plasma samples, they have a strong shielding effect on the separation and analysis of low abundance proteins derived from tissue protein release and cell destruction. While the identification of some low-concentration proteins is also of important biological significance. In our experiment, we removed the albumin high abundance protein from the plasma and laid a good foundation for follow-up experiments. On the basis of qualitative and quantitative analysis of proteomics, our measurements contained additional information. We analyzed and determined the coagulation and hemolysis events of blood samples, so as to ensure reasonable sample treatment and reliable test analysis results.

In the course of human evolution, uric acid becomes the terminal product of purine metabolism in the human body due to the mutation of the uric acid gene. In normal people, the majority of uric acid exists in the form of urate. If the body produces too much uric acid or excretes too little, it will lead to hyperuricemia. At present, the prevalence of hyperuricemia is increasing. According to a 2014 study, the adjusted prevalence of hyperuricemia among Chinese adults in 2009–2010 was 8.4–13.3% [6]. For every 60 umol/L increase in uric acid, the risk of kidney disease increased by 7 or 11% [11]. Hyperuricemia is the most important biochemical basis of gout. Recently, with the increase of social affluence and the change of diet structure, gout has become popular all over the world; the prevalence rate of gout in China is 1–3%. Gout can be characterized by gouty arthritis, gouty stones, urinary acid kidney stones, and gouty nephropathy. About one-third of patients with primary hyperuricemia and gout have renal complications. At present, there are few studies on the pathogenesis of renal disease caused by urate crystals. Although many factors related to gout have been identified, the overall mechanism of gout is extremely complicated, so specific causes of gout are still unclear. In the past few years, proteomics techniques based on mass spectrometry have made significant advances [12] and studies at the proteome level will help us to better understand the changes in the mechanisms associated with gout. Based on this, it is very important to study the difference of plasma proteins in patients with gout and even patients with gout nephropathy by using the protein identification technique of mass spectrometry, so as to explore the possible pathogenesis of gout.

In this experiment, we integrated the techniques of high abundance protein removal and proteomics analysis, combined with bioinformatics and statistical tools, developed a plasma proteome analysis strategy. Seventy-two plasma proteomes were successfully measured from 24 individuals to reveal the changes of plasma proteins in patients with gout. The proteome covers deeper—about 300 proteins per individual—and covers many clinically related proteins and inflammatory markers, as well as many other functional proteins, meeting the qualitative and quantitative requirements of plasma proteins.

Uric acid has been shown to be involved in oxidative stress through two mechanisms: NADPH oxidase activation and NO bioavailability reduction. In cells, elevated serum uric acid leads to oxidative stress, which in turn leads to the release of inflammatory factors and the activation of inflammatory pathways. In our study, 32 proteins and 10 proteins were identified differentially expressed in the normal control group and gout group, and gout and gout with renal injury, respectively. It was found that many proteins in patients with gout were changed, including complement, markers of inflammatory damage, etc., and even some proteins related to the severity of the disease were also changed, such as the up-regulation of CRP and down-regulation of Gelsolin (GSN), etc., indicating increased oxidative stress and systemic inflammatory response in patients with gout.

In addition, plasma proteins were associated with clinical biochemical markers for further study. We found that many inflammatory factors are significantly correlated with SUA, such as GSN, S100A8, S100A9, ORM2, and ANXA1.

GSN, an 80–85 kDa calcium-dependent multifunctional actin-binding protein, is a protein of the gelsolin superfamily encoded on human chromosome 9 [13–15]. GSN has three known isoforms, cytoplasmic GSN (cGSN), plasma GSN (pGSN), and gelsolin-3. pGSN is the extracellular isoform and is an abundant protein in plasma and plasma gelsolin is the main scavenger system for the toxic effects of actin filaments. Many factors, including pH, calcium levels, phosphoinositides concentration, and temperature regulate the gelsolin action [16]. A major proportion of pGSN is secreted by smooth, skeletal, and cardiac muscle cells. pGSN, composed of about 800 amino acids, is highly identical in all elements, including the structure and function among mammals. The circulating concentration of pGSN in human plasma varies from 200–300 mg/L and has been shown to exhibit a half-life of 2–3 days in humans [17].

Gelsolin is involved in the immune response and is considered an anti-inflammatory modulator. Plasma gelsolin can trigger an anti-inflammatory response by binding to the pro-inflammatory mediators; it also improves macrophages' functions via activating the nitric oxide synthase. The depletion of gelsolin causes damage in the immune cells with subsequent inflammatory mediator release. Plasma gelsolin levels' reduction was first detected in tissues damaged by trauma. Then, for GSN, it was established that the level of pGSN declines in a variety of ailments such as inflammation, type 2 diabetes, trauma, sepsis, rheumatoid arthritis, and multiple sclerosis, fibroblasts, and smooth and skeletal muscle cells [16–19]. In CKD, low pGSN levels are detectable in hemodialysed blood. This reduction is explained by many theories, such as "lower production, redistribution to the inflammation site, combined with other plasma proteins, or higher degradation" [15]. Indeed, there is a robust correlation between how much pGSN levels decrease and the probability of mortality [20]. Many studies have shown that reduced GSN levels can predict the severity of the disease and supplementing exogenous recombinant human gelsolin alleviated distress symptoms in many disease conditions [16]. The administration of recombinant plasma gelsolin to animal models reduced the inflammatory reaction [15]. This raised the question as to whether circulating pGSN levels can serve as a biomarker since pGSN levels correlate negatively with the degree of systemic inflammation and muscle wasting. However, the change of gelsolin in gout patients is still unclear. In our study, the serum uric acid level of patients with gout was significantly higher than that of the normal control group, and the serum uric acid level of patients with gout with renal injury was higher than that of patients with gout alone (although the difference did not reach statistical significance); the GSN level of gout patients was lower than that of the normal control group, ELISA further confirmed that the level of GSN gradually decreased with the normal control group, gout patients and gout patients with renal injury. It is suggested that the severity of the disease and the occurrence of complications may play a warning role. Thus, this work preliminarily investigated the change of gelsolin in gout patients. Our findings support further investigation of pGSN as an adjunctive therapy for gout.

5. Conclusions

In this study, we identified the profile of gout plasma proteome using label-free quantitative proteomics based on HPLC-MS/MS, and found 32 significantly differential proteins in the gout group, as well as 10 proteins linked with kidney damage. These abnormal proteins were characterized to biological functions by the biological tool. Further, the correlation analysis showed that five inflammatory factors (GSN, S100A8, S100A9, ORM2, and ANXA1) had a significant correlation with SUA, and GSN was further validated in clinical samples. These results suggest that hyperuricemia can induce oxidative stress and inflammatory reaction in the cells of patients with gout by promoting oxidation. Our result provides promising candidates for a biomarker for gout.

Author Contributions: L.S. and K.Z. conceived and designed the experiments. L.S. and H.D. performed the experiments. L.S., Z.G., G.Z., and K.Z. analyzed the data. L.S. and K.Z. wrote and critically revised the manuscript. All authors have read and agreed to the published version of the manuscript.

Funding: This research was supported by funding from the National Natural Science Foundation of China to K.Z. (21874100 and 22074103) and to G.Z. (21904097), Tianjin Municipal Science and Technology Commission to K.Z. (19JCZDJC35000) and the Talent Excellence Program from Tianjin Medical University to K.Z.

Institutional Review Board Statement: The study was conducted according to the guidelines of the Declaration of Helsinki, and approved by the Ethical Committee in Tianjin Institute of Endocrinology (reference number: DXBYYhMEC2018-19).

Data Availability Statement: Data are available from the authors upon the reasonable request.

Acknowledgments: We would like to thank Bei Sun (Tianjin Medical University Chu Hsien-I Memorial Hospital & Tianjin Institute of Endocrinology) for support in ELISA analysis.

Conflicts of Interest: The authors declare no conflict of interest.

References

1. C'abau, G.; Crisan, T.O.; Kluck, V.; Popp, R.A.; Joosten, L.A.B. Urate-induced immune programming: Consequences for gouty arthritis and hyperuricemia. *Immunol. Rev.* **2020**, *294*, 92–105. [CrossRef]
2. Kuriyama, S. Dotinurad: A novel selective urate reabsorption inhibitor as a future therapeutic option for hyperuricemia. *Clin Exp. Nephrol.* **2020**, *24*, 1–5. [CrossRef] [PubMed]
3. Kim, S.; Chang, Y.; Yun, K.E.; Jung, H.S.; Lee, S.J.; Shin, H.; Ryu, S. Development of Nephrolithiasis in Asymptomatic Hyperuricemia: A Cohort Study. *Am. J. Kidney Dis.* **2017**, *70*, 173–181. [CrossRef] [PubMed]
4. Benn, C.L.; Dua, P.; Gurrell, R.; Loudon, P.; Pike, A.; Storer, R.I.; Vangjeli, C. Physiology of Hyperuricemia and Urate-Lowering Treatments. *Front. Med. Lausanne* **2018**, *5*. [CrossRef] [PubMed]
5. Mena-Sanchez, G.; Babio, N.; Becerra-Tomas, N.; Martinez-Gonzalez, M.A.; Diaz-Lopez, A.; Corella, D.; Zomeno, M.D.; Romaguera, D.; Vioque, J.; Alonso-Gomez, A.M.; et al. Association between dairy product consumption and hyperuricemia in an elderly population with metabolic syndrome. *Nutr. Metab. Cardiovas* **2020**, *30*, 214–222. [CrossRef] [PubMed]
6. Punzi, L.; Scanu, A.; Galozzi, P.; Luisetto, R.; Spinella, P.; Scire, C.A.; Oliviero, F. One year in review 2020: Gout. *Clin. Exp. Rheumatol.* **2020**, *38*, 807–821. [PubMed]
7. Liu, H.; Zhang, X.M.; Wang, Y.L.; Liu, B.C. Prevalence of hyperuricemia among Chinese adults: A national cross-sectional survey using multistage, stratified sampling. *J. Nephrol* **2014**, *27*, 653–658. [CrossRef] [PubMed]
8. Singh, J.A.; Gaffo, A. Gout epidemiology and comorbidities. *Semin. Arthritis Rheu* **2020**, *50*, S11–S16. [CrossRef] [PubMed]
9. Geyer, P.E.; Albrechtsen, N.J.W.; Tyanova, S.; Grassl, N.; Iepsen, E.W.; Lundgren, J.; Madsbad, S.; Holst, J.J.; Torekov, S.S.; Mann, M. Proteomics reveals the effects of sustained weight loss on the human plasma proteome. *Mol. Syst. Biol.* **2016**, *12*. [CrossRef] [PubMed]
10. Geyer, P.E.; Holdt, L.M.; Teupser, D.; Mann, M. Revisiting biomarker discovery by plasma proteomics. *Mol. Syst. Biol.* **2017**, *13*. [CrossRef] [PubMed]
11. Weiner, D.E.; Tighiouart, H.; Elsayed, E.F.; Griffith, J.L.; Salem, D.N.; Levey, A.S. Uric acid and incident kidney disease in the community. *J. Am. Soc. Nephrol.* **2008**, *19*, 1204–1211. [CrossRef] [PubMed]
12. Geyer, P.E.; Kulak, N.A.; Pichler, G.; Holdt, L.M.; Teupser, D.; Mann, M. Plasma Proteome Profiling to Assess Human Health and Disease. *Cell Syst.* **2016**, *2*, 185–195. [CrossRef] [PubMed]
13. Martinez-Aguilar, M.M.; Aparicio-Bautista, D.I.; Ramirez-Salazar, E.G.; Reyes-Grajeda, J.P.; De la Cruz-Montoya, A.H.; Antuna-Puente, B.; Hidalgo-Bravo, A.; Rivera-Paredez, B.; Ramirez-Palacios, P.; Quiterio, M.; et al. Serum Proteomic Analysis Reveals Vitamin D-Binding Protein (VDBP) as a Potential Biomarker for Low Bone Mineral Density in Mexican Postmenopausal Women. *Nutrients* **2019**, *11*, 2853. [CrossRef] [PubMed]
14. Asare-Werehene, M.; Nakka, K.; Reunov, A.; Chiu, C.T.; Lee, W.T.; Abedini, M.R.; Wang, P.W.; Shieh, D.B.; Dilworth, F.J.; Carmona, E.; et al. The exosome-mediated autocrine and paracrine actions of plasma gelsolin in ovarian cancer chemoresistance. *Oncogene* **2020**, *39*, 1600–1616. [CrossRef] [PubMed]
15. Esawy, M.M.; Makram, W.K.; Albalat, W.; Shabana, M.A. Plasma gelsolin levels in patients with psoriatic arthritis: A possible novel marker. *Clin. Rheumatol.* **2020**, *39*, 1881–1888. [CrossRef] [PubMed]
16. Vaid, B.; Chopra, B.S.; Raut, S.; Sagar, A.; Badmalia, M.D.; Ashish; Khatri, N. Antioxidant and Wound Healing Property of Gelsolin in 3T3-L1 Cells. *Oxid. Med. Cell. Longev.* **2020**, *2020*, 4045365. [CrossRef] [PubMed]
17. Gupta, A.K.; Chopra, B.S.; Vaid, B.; Sagar, A.; Raut, S.; Badmalia, M.D.; Ashish; Khatri, N. Protective effects of gelsolin in acute pulmonary thromboembolism and thrombosis in the carotid artery of mice. *PLoS ONE* **2019**, *14*, e0215717. [CrossRef] [PubMed]
18. Asare-Werehene, M.; Communal, L.; Carmona, E.; Le, T.; Provencher, D.; Mes-Masson, A.M.; Tsang, B.K. Pre-operative Circulating Plasma Gelsolin Predicts Residual Disease and Detects Early Stage Ovarian Cancer. *Sci. Rep.* **2019**, *9*, 13924. [CrossRef] [PubMed]
19. Feldt, J.; Schicht, M.; Garreis, F.; Welss, J.; Schneider, U.W.; Paulsen, F. Structure, regulation and related diseases of the actin-binding protein gelsolin. *Expert Rev. Mol. Med.* **2019**, *20*, e7. [CrossRef] [PubMed]
20. Yang, Z.P.; Bedugnis, A.; Levinson, S.; DiNubile, M.; Stossel, T.; Lu, Q.; Kobzik, L. Delayed Administration of Recombinant Plasma Gelsolin Improves Survival in a Murine Model of Penicillin-Susceptible and Penicillin-Resistant Pneumococcal Pneumonia. *J. Infect. Dis.* **2019**, *220*, 1498–1502. [CrossRef] [PubMed]

MDPI
St. Alban-Anlage 66
4052 Basel
Switzerland
Tel. +41 61 683 77 34
Fax +41 61 302 89 18
www.mdpi.com

Separations Editorial Office
E-mail: separations@mdpi.com
www.mdpi.com/journal/separations

www.ingramcontent.com/pod-product-compliance
Lightning Source LLC
Chambersburg PA
CBHW062051220926
51850CB00010B/2733